LONDON'S LOST
POWER STATIONS
AND GASWORKS

LONDON'S LOST POWER STATIONS AND GASWORKS

BEN PEDROCHE

First published 2013

The History Press
The Mill, Brimscombe Port
Stroud, Gloucestershire, GL5 2QG
www.thehistorypress.co.uk

British Library Cataloguing in Publication Data.
A catalogue record for this book is available from the British Library.

ISBN 978 0 7524 8761 8

Typesetting and origination by The History Press
Printed in Great Britain

CONTENTS

ACKNOWLEDGEMENTS

Thank you to my wonderful wife Louise Pedroche for the never-ending love, enthusiasm and support for the writing of this book. Your patience and understanding during the long nights of writing and hectic research trips are very much appreciated. Thanks also for taking so many of the amazing photos included. Thanks to all my family and friends, and the various people and organisations who have allowed me to use archived images, especially Paul Talling, the London Transport Museum, the Museum of London, the Science Museum, the Greenwich Heritage Centre, The Enfield Society, Helmut Zozmann, Neil Clifton and many others. I owe a great debt to the authors of several books and websites included in the bibliography section. Their detailed work helped me research a subject that isn't widely covered. Lastly, thank you to everyone at The History Press for being there every step of the way.

INTRODUCTION

London is a city in a constant state of change. Always moving, always evolving, the same as it has done for centuries. It is a place at the forefront of everything ultra-modern, but it is also defined by its glorious past. Its rich history has been preserved in hundreds of buildings across the city, in particular its landmarks associated with religion, royalty, government and war. An area often overlooked by history is London's industrial past, and it is only in recent decades that great effort has been made towards the preservation of buildings that often used to dominate whole districts, and provide work for thousands of men.

The impact of London's docks and railway network are two subjects that, quite rightly, have commanded the respect of historians and the interest of many others. There are other industries, however, that seem largely forgotten, despite their legacy and the influence they have had on shaping modern London. Two of these areas in particular are London's gas manufacturing industry, and its electrical power generation industry.

Perhaps one of the reasons is that former industries like these lack much of the romance and nostalgia of others. The image of London as the world's most important port is one that's been captured many times in literature and on film. Similarly, the golden age of the railways evokes sentimental memories of a time often naively perceived to have been more simple and easy.

In reality, life in nineteenth-century London was far from simple, with millions living in poverty. Men, women and children were forced to work and live close to the so-called 'stink industries' that dominated large parts of the growing metropolis. Among the worst contributors to a poor quality of life were mills, chemical works, factories and, of course, gasworks and early power stations.

The dirty legacy of such places is probably the reason why they are often only briefly mentioned in the timeline of London's story. The manufacture of gas was a hot and hazardous job, particularly in the early years of the industry. Electric power stations were also places undesirable to most. Both relied heavily on the burning of coal in order to create the energy they needed, and therefore filled the London sky with pollutants at a time when the concept of a carbon footprint would have been difficult to explain.

Yet this was simply Victorian London at the height of its industrial heyday, where dirty and dangerous jobs were just part of normal life for thousands of workers. Every great engineering feat of the Victorian era involved men having to work in wretched and sometimes deadly conditions, from Marc and Isambard Kingdom Brunel's Thames Tunnel, to Joseph Bazalgette's sewer system and the first underground railways.

With pioneering works such as these, however, the historical focus is almost entirely on the achievements themselves, and the visionary men that created them. But to dismiss gasworks and power stations each as nothing more than simply one of many industries that came of age in the nineteenth century would be a disservice to the great innovators who built them, and the work-force that brought them to life.

Gas production quite literally enabled London to become a brighter and better developed place. As the nineteenth century drew to a close and the next one started, gas began to be superseded by electricity. This in turn gave Londoners power direct to their homes, helping the city's continued growth as the centre of the empire.

Just as with the railways and shipping companies, the firms that ran the gas manufacture and electrical supply boasted long and eloquent names that befit their importance. The men who created them and served as chairmen and directors were well-turned-out gentlemen, among some of the sharpest minds in the City of London and beyond.

The most impressive thing about power stations and gasworks was usu-ally the buildings that housed them. For gasworks, it was the huge gasholders and their frames that stood out for miles. It is only when seen from close up that their fine craftsmanship and often highly decorative features are revealed. For power stations, it was mega-sized brick structures and their chimneys that proved most awe-inspiring, especially those that once lined the banks of the Thames.

Much has fortunately been preserved for all to see, with many sites given a second life under a different purpose. The former power station at Bankside is now home to one of the world's most successful art galleries, and Battersea Power Station is a building familiar to all Londoners and many from fur-ther afield. Elsewhere across the city, huge gasholder frames still stand proud, and indeed many are still used in some capacity by today's modern gas supply companies.

Other buildings and structures still stand but their former uses have long been forgotten, like at the minimalist bar in Shoreditch that once housed electricity generators, or the rotting wooden pier supports in the Thames at Deptford that once delivered coal to one of the biggest power stations in the world. Many more buildings have simply disappeared without a trace, with nothing left behind to suggest they ever even existed.

London's Lost Power Stations and Gasworks is a collection of the finest and most important examples of each, regardless of whether they still exist today or not. It is a journey through the story behind each building and the companies that built them. It is also an exploration of the historical context of London of which they were a part, and the legacy they have left us with today.

The buildings and structures covered in the pages that follow played a vital role in the development of London as the greatest city of the world. Therefore they deserve their place in the history of London just as much as other revered structures and monuments.

The sheer size of London's disused power stations has been a source of fascination for the author for a number of years. Similarly captivating are the gritty and urban aesthetics that abandoned buildings and gasholders create.

It is an interest that began in wonder at the imposing telescopic gasholders often passed as a young boy in the author's city of birth, Nottingham. It is a passion that has continued to grow since moving to London, and part of a wider interest in the city's lost industries.

1

A NEW THIRST FOR ENERGY

London's need for power was the result of several major developments in the 1800s and early part of the twentieth century. It was perhaps the most important period in the city's long history; a time during which London grew in size and stature to become the most vital city on earth, before later being ravaged by the effects of two catastrophic world wars, only to emerge strong as always.

GAS POWER FOR LIGHTING

Gas production was introduced to London in the first two decades of the nineteenth century, and progressed during the beginning of the reign of Queen Victoria. It was an era defined by advancements in technology and science, changes in government both home and abroad, and by the fruits of the Industrial Revolution.

The British Empire grew ever bigger, with each new territory opening up greater trade routes to the rest of the world. The Port of London quickly developed into the largest collection of docks, wharves and riverside factories the world had ever seen, built to handle the huge selection of goods flowing in and out of the city. The success of the docks led to thousands of new buildings being constructed, with many villages and towns becoming absorbed into the capital as its sprawl rapidly increased outwards from the original medieval City of London. The industry that grew in and around the docks provided work for hundreds of thousands of men across London, particularly those from the East End. Many would later lose their jobs in the second half of the century as the docks began to suffer from the first signs of the major decline that would come in the decades that followed.

By the mid-1800s, London was again leading the way at the cutting edge of another major new innovation, this time on dry land. The coming of the railways had a huge impact on the city, with miles of new lines constructed that helped to connect the growing metropolis with the rest of Great Britain. All of London's most famous railway stations were built in this period, allowing for millions of people to flow in and out of the city every year.

In the early 1860s the streets of London had become such a congested mess that a new system for how people moved around the city was desperately needed. The only solution was to go underground, and it was at this time when the first lines on what would later be called the Tube were opened.

Other industries also developed across the city, including clothing manufacture, flour mills and hundreds of factories. Conditions were often appalling, with thousands of men and women forced to work in factory jobs that in many cases were comparable to slave labour.

Occurring in tandem with the establishment of London's new industries was the other major social change that defined the Victorian era; a dramatic rise in population that would see millions of new arrivals to London. The growth of industry had created new jobs for London's workforce. But the coming of the railways also made it easier than ever before for those from other parts of the surrounding counties to move to the city in order to find work. An increasing number of people also began to arrive from all over the world. Each new group of immigrants began to settle in the capital and establish their own communities, starting their own types of industry in the process. The Irish and the Jewish accounted for much of the influx of new arrivals, followed closely behind by those from Bangladesh, India and several other nations.

Other factors that contributed to the population increase included vast improvements to healthcare and medical treatment. A rise in the number of marriages led to more women giving birth, and the medical advancements helped to ensure that more babies survived their birth and early years. At the other end of the life cycle, Londoners were also beginning to live longer. This was partly the result of the better healthcare standards, but also because of the fact that the nineteenth century was a time of relative peace, with no major war on home soil, and no widespread disease outbreak.

The resulting population increase brought millions of new residents to the streets of London. The more people that arrived, the more power was needed to accommodate them. The earliest gasworks, however, produced gas primarily for commercial use only. They were used to supply gas lighting for factories and a range of other different workhouses, replacing the oil lamps of a previous generation.

Class distinction in nineteenth-century London was clear cut, with no middle ground. You were either rich, or you were poor, and most were the latter. Low income rates and the rapid increase in population meant that millions lived in poverty, squeezed into cramped housing along narrow streets. The East End in particular was infamous for its slums and the depravity that such places tended to breed.

Gas lighting for private, residential use was therefore reserved solely for the wealthy during much of the century. Nevertheless, demand did later grow for public street lighting. Its provision by several local authorities across the city helped to lower crime rates, and allowed Londoners to do more in the evening.

By the latter part of the century gas was being manufactured on a massive scale by gasworks across London. The lower prices that mass production offered meant that gas could finally become a realistic commodity for every social class, bringing it directly into people's homes for use by gas lights and for cooking.

ELECTRICITY FOR THE MASSES

The Victorian era may have been the age of gas, but the twentieth century was dominated by a new form of power. As the new century began, advances in technology and engineering allowed for electricity to emerge as a new way of powering London. Similar to gas, it was first used only for commercial purposes, or in select homes owned by the wealthy.

Shops, restaurants and theatres are just a few examples of businesses that benefited from the foresight of their owners to install electrical lighting. The extra custom it created made such businesses the envy of other traders, and it wasn't long before there was huge demand for electrical power throughout London.

What followed soon after was a desire felt by everyday people to have electricity piped directly into their homes, first for lighting, and various other appliances in later years. A mixture of private and council-owned power stations were built to meet the expected demand, although it was in fact largely the electrical companies themselves that created the demand in the first place.

Electrical power generation on a massive scale was pioneered in London. Power could now be supplied across long distances and to millions of consumers at once. It was an innovation widely promoted by each electricity company, encouraging people in the city to want it in their homes and places of work.

In the context of today's society, it is difficult to fathom the concept of electricity being anything other than a basic public service that is always there when we need it. But in its early development it was sold as a product, creating a market that grew in much the same way as the popularity of mobile phones, the internet and other global communications did in the century that followed. It was new, it was sold as a must-have commodity, and everyone ended up wanting it.

The rise of electricity did of course have a detrimental effect on gas manufacture. Electric lighting made the gas lights of old seem archaic in comparison, and it led to serious decline for many gas companies. London's gasworks still produced a product that was in demand for cooking and heating, although this would begin to disappear by the 1960s.

POWER FOR TRANSPORT

Another increase in demand for electricity came from the continued success of London's transport system. In the late nineteenth century and early part of the twentieth, electric trains rapidly replaced steam on the Underground network. More and more people were using the Tube every day, resulting in a higher frequency of trains per hour being needed. This called for high amounts of electric power, and with no National Grid at this stage, it was left to the responsibility of the railway companies to generate electricity themselves.

The initial solution was a number of small power stations owned by individual lines, but as passenger numbers increased and companies amalgamated into one network, it was clear that a more suitable long-term plan was required. The outcome was a series of large power stations constructed specifically to power the network, with extra capacity left over to supply new lines like the Victoria and Jubilee that would arrive later. Elsewhere, the city's tram network also needed powering in the early decades of the twentieth century. This was managed in much the same way as the Underground, with tramway operators having their own small generating plants, before later being powered by the dedicated large power stations. London's main-line railways also later made the switch from steam to electric traction, which again demanded power from a series of purpose-built power stations across the city.

The success of the Underground network was itself a by-product of London's continued growth during the last century, both in population and physical size. The first half of the twentieth century was characterised by the devastating impact of two world wars. Yet population continued to rise regardless, especially in the period between wars, when London spread out in every direction, causing its number of residents to swell to more than eight million.

The mass of people, and the products and services they demanded, meant that London was able to avoid being impacted too greatly by the global economic downturn of the 1930s. Rather than suffer decline, many industries continued to grow, among them the electricity supply companies.

By the 1940s, the landscape of London was dominated by large power stations. The British Empire had perhaps lost much of its power, and other cities around the world may have taken away some of London's status as the centre of the world, but it was still a city on the rise. As technology, youth culture, arts and entertainment, tourism and commerce all continued to grow, so did the need for electrical power. It is a demand that has never eased since. Today, London still requires huge amounts of gas and electricity. Gasworks have been replaced by supply from the North Sea and beyond. The major power stations have all but gone, replaced by more efficient and environmentally friendly plants. But it is still a city hungry for power, a necessity not likely to ever disappear.

2

BIRTH OF THE INDUSTRY & THE MAJOR PLAYERS

With London's ever-growing need for power in the nineteenth and twentieth centuries, it didn't take long for the wheels of commerce to start turning in unison with the mechanical and industrial developments they began. Demand was high, and it was a demand that needed to be met. Although London's electricity supply, gas manufacture and other such power-related industries developed at different times and in different ways, the principle was always the same. Scores of different companies started to appear almost overnight, creating fierce competition within each sector, and often mass confusion for the early consumers.

As is usually the case in business, however, many firms disappeared just as quickly as they appeared, allowing a small group of large companies to take control, absorbing everything in their path towards the top. The birth of the gas industry in particular is a tale littered with faded memories of lost companies that were forced to amalgamate with a handful of major businesses. Other companies managed to dominate their industry from the very start, having become established so early that no one else ever really stood a chance of competing against them.

This chapter explores how each power industry emerged, and the major players within them. Many of the power stations, gasworks and other buildings discussed in this chapter still exist, and will be explored in full detail in later sections of the book. Further information about how each power system actually works will also be covered later.

GAS PRODUCTION ARRIVES IN LONDON

Electricity rapidly earned its place as the leading power industry in the latter part of the nineteenth century and into the twentieth. But before that there was gas manufacture, a genuine stink industry that came of age in Victorian London.

Its primary use in the 1800s was to provide lighting, specifically by powering gas lights. Although not capable of achieving the same brightness as the electrical lights that would later replace them, gas lights were a vast improvement on the candles and oil lamps that came before them. Instances exist of gas being recommended as a way of powering lights as far back as in the 1680s, suggesting that the basic principle was far from new. But it was only in the early part of the nineteenth century that sufficient advances in chemistry meant that what was once merely theoretical could now become a reality.

There was little demand for public gas supply at first. The early companies that began to operate instead focused their attention on providing lighting for factories and mills. The appeal was simple: gas lights were cheaper, and more powerful and efficient than oil lamps or candles. The lack of a naked flame also provided a safer environment, especially for those factories manufacturing flammable products.

Taking note of how water was already being supplied directly to buildings and houses by way of mains built below streets, the potential for gas to be delivered to consumers in the same way was soon realised. It wasn't long before the focus of attention shifted therefore from industry to the public domain.

Although London was often surprisingly lacking compared to other parts of Britain when it came to capitalising on new industrial advancements, it was in fact at the centre of the early gas industry. And so it was in London, in Westminster to be exact, where the world's first public gas supply company was established in 1812 by royal charter. It was called the Gas Light & Coke Company and, quite remarkably, managed to retain its position as the biggest gas producing firm in London until it was nationalised more than 130 years later.

The company owes much of its success and ambition to German inventor Frederick Albert Winsor. Something of a fantasist and unabashed self-promoter, Winsor came to London in 1803 to generate interest in his pioneering work relating to gas lights and their potential for public lighting. Yet even at this point in time, at the very beginning of the century, the concept was not entirely new. It was an idea already being developed in Paris, but also closer to home in Birmingham.

The advancements in the West Midlands were thanks to Scottish engineer William Murdoch. Building on the knowledge gained from reports of wealthy businessmen who had used gas to light their homes, it was Murdoch who recognised the potential for gas lighting to be used in factories. Murdoch was employed at the time by Boulton & Watt, a company whose steam engines had played a significant role in the Industrial Revolution. As a firm already renowned for its innovation, the chance to be part of a new scientific breakthrough made it easy for Murdoch to gain permission to experiment with lighting at their factory in Birmingham.

The experiments proved successful, and by 1803 gas lighting at the factory became a permanent fixture. This was followed by similar installations at various other factories. Murdoch was doomed to become merely a footnote in the history of gas supply, however, thanks to his short-sighted insistence that gas lights would only ever be suitable for industry.

Winsor on the other hand did see its public supply potential, and set about proving it with a series of early lectures and demonstrations, most of which failed to impress potential investors for his proposed National Heat and Light Company. It was a PR campaign that Winsor had already failed at in Paris, but his enthusiasm and persistence eventually started to pay off in London.

The huge potential of Winsor's idea for using mains to supply gas to the public could no longer be ignored. By 1807 a consortium of investors began a series of moves towards shaping his ideas into a serious business model. The first task was to gather support in Parliament, an exercise that was met almost immediately with opposition. Although the consortium claimed that the new proposed company – now set to be called the New Patriotic Imperial and National Light and Heat Company – was being founded on the notion that it would provide a much-needed public service, the detractors were quick to point out that a lack of competition would lead to a monopoly. One individual figure opposing the new company was Murdoch, a man perhaps more than a little bitter that his early innovation was now on the verge of major financial success without his involvement.

Whatever the motives of those opposing Winsor, the various objections raised were enough to ensure that the company didn't obtain its charter.

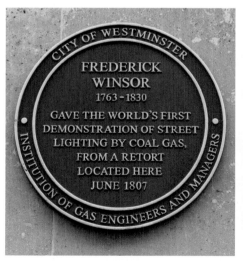

Plaque commemorating Frederick Winsor's early gas demonstrations.

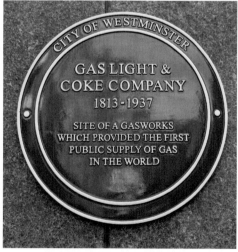

The Gas Light & Coke Company's first gasworks in Westminster.

Other unsuccessful attempts followed, but the consortium did finally succeed in gaining their royal charter in 1812. Now to be known as the Gas Light & Coke Company, the original plan for a national gas supply had been down-sized to London only, but it was still enough to lay claim to being the first company of its type the world had ever seen.

The company opened its head office and first gasworks on what is now Great Peter Street in Westminster. Keen to widen their reach as quickly as possible, the Gas Light later opened two further gasworks in Shoreditch, noto-rious in the nineteenth century as being one of the most crime-infested areas of London. The first was located on Curtain Road, while the second was on Brick Lane, built on the site of an old brewery. Neither proved to be of much use, primarily because each was located too far from either a canal or a river. A gasworks needed a constant supply of coal, and the most cost-effective method of delivering this was via barge. The lack of waterways alongside either of the Shoreditch works meant that instead the company had to pay the higher costs involved in using trucks to deliver its coal supply.

The rising demand for gas lights also meant that the two early works soon became inadequate, having little room for expansion. It wasn't until much later in the century that the company built a new and improved works. For now though, the Gas Light was firmly established as the dominant force in a new but rapidly growing industry, and within a few years it began to run at a significant profit.

Frederick Winsor was kept as an employee, and served as the public face of the company during its fight to obtain its charter. It was ultimately his lack of business acumen that led to him playing only a minor role in the Gas Light's long-term success. He later fought to be recognised officially as the creator of the company, and therefore be compensated accordingly with a share of its profits. His demands were unsuccessful, with the company going to great lengths to make it clear that their operation was different enough to Winsor's original concept for him to make any serious claim. Winsor later returned to France to try to set up a similar business in Paris, but this only led to further failure. He died soon after in 1830.

Today, Winsor is still remembered as the company's founder, and there is a monument dedicated to him in the grounds of Kensal Green Cemetery. In addition, a road close to the Gas Light's largest works was also named after him.

The success of the Gas Light & Coke Company from 1812 onwards led to a rush of other companies entering the market, and by the mid-1830s London had hundreds of public gas mains in operation. Unsurprisingly, competition between rival companies was fierce. It draws comparison with the gas and electricity supply industry of today, where consumers are constantly being encouraged to switch to a different provider.

But while today the methods by which competitors attempt to outdo each other usually involve nothing more sinister than undercutting prices, the tactics used in the nineteenth century were far more dirty and underhand. It was not uncommon for workers to be sent out to sabotage the gas mains laid by rival companies, and even steal their main's connections. Early consumers may have been able to reap the benefits of such actions, but they had a negative impact on gas company employees. If working in one of London's stink industries wasn't already hard enough, wage reductions and increased danger must have added significantly to the low morale and poor working conditions. Overheads, wage bills and safety-related maintenance were all costly activities for a company willing to cut corners in order to offset profit reductions caused by having to lower their prices to beat competitors.

It was clear that change was necessary, not only to improve the inner workings of the industry itself, but also to try and balance the success of the Gas Light with a genuine rival. It was still the most powerful company in the gas industry, regardless of how many smaller firms were out to take the throne.

It was a situation that played directly into the hands of the early detractors and their fears about a monopoly, but there was also now a second wave of men protesting at the way the industry was progressing. One of them was a Stepney businessman named Charles Hunt, who in 1837 spearheaded the creation of the Commercial Gas Light & Coke Company. It was an ambitious new firm that would give the original Gas Light & Coke Company some real competition. Hunt's manifesto was to build a company that would be owned and run by consumers, supplying a better and cheaper service than its competitors. Various local businessmen were encouraged to invest and purchase shares, and by 1839 the company had secured enough financial backing to open its own gasworks at Stepney.

The Commercial Company's strategy was to focus its attention on large areas of east London not already controlled by the Gas Light, including parts of Wapping, Poplar, Limehouse and Shadwell. Several of these were, however, districts already being supplied by smaller companies, most of which helped to ensure that the Commercial Company was given a harsh reception. In a return to the same skulduggery that was rife in the early days of the gas industry a decade before, these local companies took to digging up expensive new mains built by the Commercial Company, in particular in Mile End.

The fear was that a large company like this would take all of their business. It was in fact a reasonable fear to have, and one being felt perhaps most by the Gas Light. They could now see a serious contender on the horizon for the first time in their history.

The growing resentment towards the Commercial Company was further exacerbated when the firm achieved great financial success under the leader-

ship of the new chairman, Charles Salisbury Butler. His foresight in convincing the board that the company should expand its reach to Whitechapel,
Spitalfields and Bethnal Green proved to be a positive move, and enabled the
firm to weather its early financial struggles.

The initial reaction from the Gas Light, and indeed various other small
companies who at first had set out to ruin the Commercial Company, was
to propose ways for each company to work together. The reasoning was
simple: 'You stay away from our territory, we'll stay away from yours, and
they'll be enough consumers for everyone to make a profit.' The Commercial
Company's response was to refuse any such alliances, no doubt acutely aware
of the sense of desperation pervading the actions of their rivals.

There were, however, many occasions when the various competing gas
companies did unite in order to protect their shared interests. Strike action by
gas workers was a frequent issue, and rival firms would often lend their own
workforce to fill the void left by the men on strike elsewhere. Strikes, such as
those that took hold at gasworks and other industrial workplaces in the 1860s
and '70s, could often turn violent, with workers sometimes facing criminal
charges. This meant that when workers were sent to help out at other works,
the official line was that they were there simply to fulfil regular duties. But off
the record, they were often there as extra muscle.

It would have been hard to deny this sinister pretence as the real reason for
rival workers being deployed on one particular night in August 1850. It is a
date that marks the darkest hour in the history of the Commercial Company
and the nineteenth-century gas industry as a whole, and was given the nickname the Battle of Bow Bridge.

A new gas company known as the Great Central Gas Consumers Company
had arrived on the scene in 1847 with a view to becoming a serious threat to
the dominance of both the Gas Light and Commercial Company. Reminiscent
of the founding principles under which the Commercial Company itself was
created – many of which had long since been eroded – the Great Central
aimed to build a company that would provide gas at a lower and fairer price,
whilst still ensuring a healthy return on investment for its shareholders. Of
particular frustration for the Commercial Company was the fact that their
new rival decided to open a new gasworks at Bow Common, right in the
heart of Commercial Company territory.

Similar to the days of sabotage that the Commercial Company had itself
suffered in the early years, they deployed a team of 'workers' from their own
gasworks and those of the Gas Light one evening in early August to disrupt
workers from the Great Central as they tried to construct a new mains pipe
across Bow Bridge. They succeeded in driving away the workmen, and pro-
ceeded to construct their own mains pipe across the same bridge. It was a

main the Commercial Company did not need, and was instead built entirely for the purpose of ensuring their rival would have no room left on the bridge for a main of their own. The men then went a step further by creating a barrier across the bridge.

What they didn't account for was the size of the Great Central's defences, and within hours a 300-strong army of men arrived to overpower the Commercial Company's workers. They stormed the barricade, destroyed the new main and replaced it with their own. By the following day there were several men injured from fighting, and others arrested and remanded in custody for violent behaviour.

The overall success of the Commercial Company had a profound effect on the industry. But in reality the Gas Light remained as the market leader, and steadily increased its power by amalgamating with many of its smaller rivals. Every existing company it acquired gave the Gas Light not only increased market share but also more and more gasworks. By 1930 the company was supplying gas to huge parts of London and its surrounding districts. More than thirty different companies were absorbed by the firm between 1812–1932, giving them coverage that ran as far as Staines, Brentford, Southend-on-Sea and Barking.

The Commercial Company followed suit with a series of amalgamations of its own, including the strategic purchase of the British Company. The purchased company had previously spent significant time and resources years earlier on trying to ruin the business of their now owners.

The landscape of the industry changed again in 1868 when new legislation was passed which meant fines could be charged to companies who failed to supply gas of a high enough quality to power gas lamps to their full capacity. The result was that several companies had to modernise their existing gasworks so that they could meet the required levels of capacity, or simply build new, larger works instead.

The latter option proved to be the most viable long-term solution for the major companies, and it sparked a new era of mega-sized gasworks that soon became a common feature across London. The Commercial Company expanded its works at Stepney, and the Gas Light opened a new site in East Ham. Known as Beckton Gasworks, it was one of the world's biggest industrial sites on its opening in 1870.

Another company to open a huge gasworks in the 1870s was the Imperial Gas Light & Coke Company, who began construction of a new works at Bromley-by-Bow, complete with no fewer than seven gasometers.

The Imperial was in fact a significant player in its own right. It was founded in 1821 and soon began to gain control of significant parts of north, east and west London, with considerable works at Shoreditch, Kings Cross, Fulham and the new site at Bromley-by-Bow. A key figure in its success was Consultant Engineer Samuel Clegg, who had long been associated with the gas industry. His credentials included assisting William Murdoch with his early gas light experiments at the Boulton & Watt factory. He also acted as engineer to the Gas Light during its early years. This connection with the largest company of all would come full circle in 1876 when the Imperial became one of the many companies absorbed by the Gas Light. It was a major factor in the continued dominance of the company, giving them almost total coverage of central London and its vicinity.

The South Metropolitan Gas Company was another thriving business that built a large gasworks in the latter part of the century. It was a company in good health by 1879, boosted by recent amalgamations with the Phoenix Company and the Surrey Gas Consumers Company, two firms with lucrative coverage of London districts south of the Thames. Their major new works opened at East Greenwich in 1886, on the site of what is now the O_2 Arena. It would prove to be the last great gasworks constructed in London. The South Metropolitan continued its success into the next decade, and would later come close to purchasing the Gas Light. The deal never materialised, but the South Metropolitan did later become a major shareholder in the Commercial Company. The Gas Light had itself made several offers to amalgamate with the Commercial Company between 1873 and 1883, but was rejected each time.

By the turn of the century the gas industry was being hit hard by the growing success of electrical power, with further decline taking effect after the discovery of natural gas in the North Sea. Chapter 6 will discuss the decline of the industry and its major gasworks in further detail.

ELECTRICITY TAKES OVER

In the twenty-first century, when our entire lives revolve around technology, it is hard to imagine not having electricity when we want it. But in the latter part of the nineteenth century there were no appliances, and the electricity supply industry existed almost solely for the purpose of supplying lighting to those who demanded it.

Although this demand would later grow and indeed be met on a huge scale between the 1880s until the 1940s – with mega-sized power stations being built across London – the city's power industry actually began under far less spectacular circumstances. The story begins with an artist named Coutts Lindsay, who opened an art gallery on Bond Street in 1877. Named the Grosvenor Gallery, it was opened to showcase the work of new artists whose work was seen as too radical for the mainstream galleries.

When it was decided that electrical lighting should be added to the venue in 1883, the entrepreneurial Lindsay simply purchased the machinery needed to generate electricity himself. It didn't take long for the owners of other businesses in the Bond Street area to also want electrical lighting, and so Lindsay obliged by expanding his basic set-up to supply various other firms with power at an affordable price, via overhead cables. Within two years the supply business was so successful that Lindsay expanded again with the opening of the Grosvenor Power Station, under his new company name of Sir Coutts Lindsay Co. Ltd.

This small but significant power station continued to provide a public power supply until 1887, widening its reach along the way. When a large new power station was opened at Deptford in the same year, the site at Bond Street was downgraded to a substation, which can still be seen today on Bloomfield Place, just off what is now New Bond Street.

Around the same time as the success of the Grosvenor Gallery, a similar operation was also gaining popularity in the theatre district. After inheriting the family business from their father, Swiss brothers John Maria and Rocco Joseph Stefano Gatti had built up a successful empire that included owner-ship of restaurants and theatres, including the famous Adelphi. Just as with Lindsay's art gallery, the installation of a small generating station, built in order to power electric lighting, soon attracted the attention of other local businesses. The Gatti brothers founded the Charing Cross and Strand Electrical Supply Company to meet the demand, and were soon selling the use of its power source to large parts of London's West End.

Also operating by 1883 was a small electricity supply close to Farringdon, in the City of London. It was built by the Edison Electrical Company, owned by famed American inventor Thomas Edison, especially to provide power for street lighting along the Holborn Viaduct. This decorative bridge across the former valley of the River Fleet had been opened with much fanfare by Queen Victoria herself. It was therefore a suitable London landmark on which to experiment with electrical street lights. Although short-lived, the generating equipment's success in providing lighting not only to the viaduct but also several houses nearby meant that Edison's building at 57 Holborn Viaduct had in effect become the world's first public power station.

The move towards a mass public electrical supply came with powers granted to local councils – and other such authorities, namely parishes – under the Electrical Lighting Act of 1882. It awarded several London districts the licence to open their own power stations. The Act also opened up the opportunity for private companies to construct and operate their own power stations, but there were strict conditions attached. Any company looking to start up would first have to gain the approval of the local authority under

Grosvenor Gallery substation, near Bond Street.

which the proposed station fell. In addition, the local authority was given the option to purchase the private company's operation after a certain number of years. It was a set of rules that would ultimately be a huge step backwards for the budding industry, as unsurprisingly, few investors were willing to create a company that would be subjected to such rigid restrictions. It also led to a huge lack of consistency of delivery systems, with some power stations opting for a direct current transmission, while others preferred to use alternating current. Some companies delivered power via overhead cable, while others used cables under roads.

Another huge downside to the various local-authority-owned power stations was that they only served the people who lived in that area. It is of course a reasonable concept – why supply electrical power to people who don't live in your area? But it hampered development of what, in hindsight, London really needed; a connected network that covered the entire metropolis. In fairness to the local authorities, however, the inconsistency across different power supplies was a technological issue just as much as it was a political one.

The early breed of power stations were simply not powerful enough to supply to large areas, but that changed in 1891 thanks to advancements made by the engineer Sebastian de Ferranti. His work in the field of electrical power generation led to him being commissioned to build a huge power station at Deptford, which became the blueprint for every large-scale station that followed.

This new type of plant meant that supply from one single station could now reach far beyond localised boundaries. The advancements in technology also coincided with a series of amendments to many of the strict terms and conditions of the Act of Parliament. The changes now allowed for private companies to take full advantage of the extended range that new stations like the one at Deptford could have, and to supply power commercially to areas outside the jurisdiction of the local authority under which they fell. It was at this point that the major industry players quickly began to emerge.

One of the first companies to make an impact was the London Electric Supply Corporation, founded in 1887 specifically to take ownership of the fledging power station at the Grosvenor Gallery. The company later expanded, and it was they who commissioned Ferranti to build the new station at Deptford.

By the late 1890s the County of London Electrical Supply Company Ltd was also becoming a dominant force in the new industry. Having already run modest but successful power stations at City Road in Islington, and south of the river at Wandsworth, the company followed the example set by Deptford by constructing a large power station of their own at Barking, which opened in 1920.

It soon became clear that mega-sized power stations were the way forward, which made for a gloomy outlook for several of the smaller stations across London. Many were acquired by the dominant companies and converted into substations that were fed power by the large stations. The smaller power companies not directly owned by one of the majors began to reach the conclusion that it made better business sense to have the power they were selling generated and distributed to them by the larger companies. This meant they no longer had to cover the high costs of actually generating the power themselves.

Despite the proliferation of the larger companies and the amalgamation of various smaller ones, several local councils built their own large power stations, including one in the Sands End area of Fulham, and another at West Ham, located close to Canning Town.

Further legislation changes and parliamentary acts paved the way in 1926 for the formation of an organisation that became the absolute major player in the industry. It was given the appropriate and all-encompassing name of the London Power Company (LPC), and it quickly gained ownership of ten existing companies – including the London Electric Supply Corporation – allowing them to inherit several power stations in the process. Some of these were subsequently closed down, but most were updated with new machinery.

In the case of Deptford, the LPC built an entire new station in 1929 alongside the original, the construction of which was overseen by newly appointed Engineer in Chief Leonard Pearce. He later played a significant role in the company's most ambitious project of all; the building of Battersea Power Station, which opened in 1933. The super-sized facility served as the pinnacle of London's electrical supply industry, and since its closure has become one of the most iconic buildings in Britain.

It is important to note that the parliamentary Act that facilitated the creation of the London Power Company had as one of its main objectives the unification of each electrical system. Decades of individual companies entering the market had led to a number of differences across issues such as type of current (alternating versus direct) and the use of either overhead or buried wires. The latter required having to gain permission to dig up roads; which was often hard for private companies to obtain. The LPC was tasked with creating one system, with a view to creating a fully connected network. By the 1950s, plans towards developing what would later become the National Grid were put in place, which led to a period of rapid decline for London's power stations – chapter 6 contains more details on why most of the city's stations closed down.

THE RAILWAY GOES UNDERGROUND

Perhaps surprisingly, several of the major companies in the history of London's electrical power industry weren't even power companies at all. They were in fact transport organisations, many of which introduced electrical power across the city's underground and main-line railways. Each played a significant role in creating what today is the Transport for London network, with two companies in particular driving huge progress.

The history of underground rail in London is one that has already been told in countless other books, but its significance as a supplier of power is often only briefly covered. In actual fact, the Tube network of the early twentieth century included a complex system of electrical substations, and more significantly, independently owned power stations. The stations were of course built for the sole purpose of powering the railway lines owned by the companies that built them (as opposed to a public supply), yet with millions of Londoners using the Tube, it was one of the most significant power networks the city had ever seen.

The demand for electrical power on the underground railways hadn't yet been realised when the first line was opened in 1863 by the Metropolitan Railway (MR). The line ran from what is now Paddington, though originally known on opening as Paddington (Bishops Road), in central London to Farringdon (opened as Farringdon Street) in the City. As with all non-subterranean railways of the time, the trains themselves ran on steam, and continued to do so until the opening of the City & South London Railway (C&SLR).

The problem with steam trains was that they were loud and dirty, bellowing sulphur and carbon dioxide emissions as they travelled along the tracks. Out in the open this was rarely an issue, as the smoke was simply released into the air. Take the same trains into a network of long tunnels however, and the issue of smoke fumes becomes huge. From an infrastructure perspective, it was a problem that the engineers of the Metropolitan Railway had been acutely aware of from the very beginning. Their solution was to build vents and short sections of open tunnel, where smoke could be dispersed into the open. It was a process that was fairly easy to achieve, in large part because all of the early underground lines were constructed using the cut-and-cover method. This meant that the tunnels were usually close to the surface (see the following section on hydraulic power for more on the early cut-and-cover lines).

But as more and more Londoners began to use the new underground railways, it became clear that the steam was fast becoming an issue that was affecting business, and even effecting public health. Accounts of hellish, smoke-filled platforms and trains were not uncommon, and passengers regularly complained of coughs and sore eyes. They were symptoms that led to several London chemists prescribing an elixir that was nicknamed 'Metropolitan Mixture'. The health issue was played down by the railway companies, and even promoted by the management as being beneficial to good health (it was suggested that smoke from steam engine emission contained acids that could help alleviate illnesses).

It became a problem that was impossible to ignore, and grew worse in 1868 when the newly formed Metropolitan District Railway (MDR) began operating its own trains as part of the Inner Circle, which today forms much of the Circle line.

The answer was electrical power, a new innovation first introduced by the C&SLR in 1890. As the first deep-level Tube line (i.e. not constructed using the cut-and-cover method), providing outlet vents for steam trains to release smoke was no longer now an option, as the new tunnels were to be constructed far deeper below the surface.

The original plan was for C&SLR trains to be powered by cable. But at the eleventh hour it was announced instead that the line would be electrified; a decision likely inspired by the simple yet pioneering electric traction railways in Europe, and the Volks Electric Railway in Brighton.

However, with no public electricity network to tap into, the C&SLR had to create its own power in order to operate. A generating station was therefore constructed at the terminus of the line at Stockwell, making the company one of the early major players in the power industry.

From the initial success of the C&SLR line from Stockwell to King William Street, it was evident that electrified trains were the future of underground railways in London. This new system of power was cheaper and better for business, as it meant train services could become faster and more efficient for consumers. The removal of smoke also solved the health and comfort problems faced by passengers, which itself led to increased revenue.

It is hardly surprising, therefore, that all Tube lines built after the C&SLR were electrified. The Metropolitan Railway and Metropolitan District Railway also converted their original lines to electricity, although steam was still used for maintenance trains on the Met line well into the twentieth century.

Just as the C&SLR had to provide its own electrical power via the dedicated station at Stockwell, the move to electric power meant that the other underground companies had to do the same. But while the generating plant at Stockwell was a fairly modest affair, the larger companies had the success and ego to construct their own huge power stations that matched those being built outside of the transport system.

There was Neasden Power Station in north London, which was built by the Metropolitan Railway to power their entire line after the switch from steam power to electricity was complete. An even grander power station was constructed in west London as part of an ambitious expansion plan engineered by Charles Yerkes, one of the most important men in the history of London's underground railway.

An American businessman with something of a chequered past, Yerkes had already made an impact on urban transit systems in Chicago when he arrived in London at the turn of the twentieth century. He founded the Underground Electric Railways Company of London Limited (UERL), who took ownership of the Metropolitan District Railway soon after. The company later acquired underground lines that today form the Piccadilly, Bakerloo and Northern line, making the UERL the dominant force in London's subterranean railway, until the entire network was later consolidated.

Ignoring the alternating current favoured by the Metropolitan District Railway – based on testing which the company had carried out, together with the Met, along part of the Circle line before its acquisition by the UERL – Yerkes instead decided his electrified trains would run on a direct current system.

The fact that the Metropolitan Railway was so keen to use the AC system led to legal action against Yerkes' decision. The court, however, ruled in favour of the UERL. The controversial decision to allow different electrical systems on the same network was just one of several inconsistencies that plagued the

London Underground when it was unified as one company. It is a legacy that still has an impact today.

None of this mattered to Yerkes, who ordered the construction of a huge power station at Lots Road in Chelsea. This was used to power the MDR and later every other line under the ownership of Yerkes' company. When the underground was consolidated in the 1930s under the London Transport Passenger Board, the rival power stations at Neasden and Lots Road joined forces to power the entire network, assisted by smaller substations like the ones still in use today at Northfields, Wood Green and various other stations.

The Underground wasn't the only form of electric transport that needed powering however. By the early twentieth century London had an extensive tram system that covered huge parts of the metropolis in every direction. It was a mode of passenger travel that had been progressing ever since George Shillibeer had introduced the first omnibus route in 1829. By the late nineteenth century steam-powered trams had started to dominate, but by 1901 electric traction rapidly replaced all previous systems.

It was during this time that the inevitable small group of major companies began to emerge, namely London United Tramways (LUT), Metropolitan Electric Tramways (MET) and London County Council Tramways (LCCT). Similar to what was happening below ground, the various competing companies experimented with different forms of traction. The LUT, for instance, used a network of overhead cables, while the LCCT pioneered the use of conductor rails placed in conduits along the road.

A source of power was required regardless of which system was adopted. This forced the major tram companies to build their own power generating stations. The LUT constructed several small plants, the most significant one being located at its headquarters and depot in Chiswick. The LCCT, meanwhile, went for something on a far bigger scale by building a huge power station in Greenwich, which was also later used alongside Lots Road to power the London Underground.

As if to further underline the importance of Charles Yerkes as a dominant figure in London's transport history, and by extension his role in its electrical power industry, he was also involved in the tram network. His UERL company became joint owner of London United Tramways and Metropolitan Electric Tramways in 1913, before the entire company was brought under the control of the London Transport Passenger Board in the 1930s.

London's main-line railways were among the last to make the conversion to electrical power, although it is not hard to understand why. With coal supply in abundance and no concerns over public health or comfort regarding smoke emission, steam continued to prevail on the city's open air railways.

But the increased power and efficiency that electrification offered began to be an opportunity too good to miss. By the first decade of the twentieth century several miles of London's railways had been converted to electric power, most significantly along lines owned by the London & North Western Railway (LNWR).

Mirroring the inconsistencies on the Tube and tram networks, different electrical current systems were employed by different rail companies. Some utilised overhead wires, others a third-rail system, and others a fourth-rail variation.

Power itself was generated at a number of small-scale stations, the biggest of which was one owned by the LNWR at Stonebridge Park. These power stations tended to be fairly low key in design when compared to the likes of Lots Road and other such transport sites. Considering how grand the mainline London termini of each major railway company tended to be, it is likely that owners didn't see much value in building a power station in anything other than a functional style, as it would not be something seen by customers.

POWER FROM WATER

The origins of London's hydraulic power supply industry began with two men in particular, neither of whom were from London. The first was Joseph Bramah, an inventor from Yorkshire, who in 1812 pioneered the idea of a power network based on his invention of the hydraulic press.

The other key figure was solicitor-turned-industrialist William Armstrong from Newcastle. By 1845 Armstrong had further developed Bramah's work, and begun to see its potential as an idea that had commercial appeal. Drawing comparisons with the early gas industry, Armstrong turned to heavy industry as an area where a hydraulic power network could best be applied. In 1845 he successfully installed a hydraulic-powered crane on the docks of the River Tyne in Newcastle. The prospect of more efficient and cheaper cranes had obvious appeal to docking companies and railways, and it wasn't long before Armstrong's new type of crane was in heavy demand. He founded W.G. Armstrong & Company in 1847, and by the early 1860s the firm had grown into a major operation that employed over 3,000 men.

With the use of hydraulic power now well established for industrial purposes, the inevitable next step was to develop ways for it to be applied to public use via mains pipes that could supply power directly to buildings. It was a concept pioneered largely by London-born entrepreneur Edward Bayzard Ellington, who was influential in the passing of an Act of Parliament that allowed for the supply of water at high pressure to cities and towns. Ellington founded the Hull Hydraulic Power Company in 1872, which began supplying Kingston upon Hull in 1876, using water from the Humber.

It proved to be a success, and Ellington arrived back in London in 1882 to act as engineer for the General Hydraulic Power Company. The new firm merged a year later with a similar outfit known as the Wharves and Warehouses Steam Power and Hydraulic Pressure Company. The combined new firm became known as the London Hydraulic Power Company, with offices on Grosvenor Road in Pimlico. The company was the result of an 1883 Act of Parliament allowing for a system of mains to be built under London's streets for a pressurised power supply.

It was to become the dominant and essentially the only company of its type in London, with hydraulic pumping stations located at Blackfriars, Wapping, Rotherhithe and Clerkenwell. It owned more than 180 miles of mains pipes by 1927, supplying an incredible 1,650 million gallons of pressurised water a year in every direction across London. The power was used in factories, docks, train yards, hotels, offices and many more public buildings. The network operated until closure in 1977, following years of decline due to competition from the electricity industry.

3

THE GREAT CATHEDRALS OF POWER

Many of London's old power stations have disappeared without trace. Others have been demolished and redeveloped, leaving only small clues to suggest that they ever existed. There are several still standing, however, including some of the finest architecture anywhere in London. Some are huge, mega-sized buildings that have struggled to find a new use since closure, while others have been repurposed with great success.

Regardless of whether they still exist or not, all of the buildings listed here are memories of a time when one of the most historic cities in the world was dominated by power stations that brought electricity to millions. With smoke pouring out of tall chimneys as coal delivered by boat or train was burnt all day long, these loud and dirty buildings once lined the banks of the Thames and beyond. It was an era completely removed from the luxury apartments and entertainment complexes that run alongside the river today, signalling the death of heavy industry in the nation's capital.

It is clear that London is undoubtedly better off without the stink industries of the Victorian age and the decades that followed. But the legacy of what came from the foresight of pioneers like Sebastian de Ferranti, Coutts Lindsay, Charles Yerkes and Giles Gilbert Scott still has an impact today, and it is their work that has helped maintain London's status as one of the world's most important places.

What follows is a look at some of the most important power stations in London's electrical supply history. First, however, a brief look at how a power station works. Note that the process described here is based on a coal-fired power station, which most of London's lost stations were. The process differs slightly for other types of fossil fuel, such as oil or gas.

Coal is delivered to a power station and burned in huge boilers. This heats water, which in turn creates steam. The steam can then be used to provide power to drive turbines, which then supply energy to alternators that generate an electrical current. The electrical supply can then be provided to consumers via overhead wire to substations, and then to individual buildings through electricity mains.

POWER STATIONS FOR PUBLIC SUPPLY

Deptford

Mega-sized power stations were at one time a common site along the banks of the river Thames. Some remain, like Battersea and Bankside, but most have been lost, including the one that became the benchmark for them all. On its opening in 1891, Deptford Power Station was the largest electricity generating site in the whole world, built on a scale that had never before been seen. It was the vision of engineer Sebastian de Ferranti, who helped create the London Electrical Supply Corporation (LESC) in order to make his planned power station a reality. The company was the successor to the Sir Coutts Lindsay Company discussed earlier, set up by the owners of the ground-breaking Grosvenor Gallery in order to expand their business.

Ferranti's idea was simple. Instead of many small power stations supplying electricity to many different areas, what London needed was one huge station that could supply to thousands. Investors were sceptical, and it was decided that such an experimental undertaking was too dangerous to be built within central London, for fear of what may happen if disaster struck. Instead,

Deptford Power Station. (Image used courtesy of the Greenwich Heritage Centre)

a plot of land was chosen in south-east London, close to where the River Ravensbourne meets the Thames at Deptford Creek.

Late nineteenth-century Deptford was a place suffering from major decline after its once famous docks had become obsolete. The prospect of a major new power station, and the boost to the local economy and employment levels it would bring, was therefore welcomed with open arms by local officials. The chosen site also suited Ferranti, who was keen to make use of the Thames as a source of water that he knew would be needed for the power station's large cooling towers.

Construction work began in 1888, with even the building itself designed by Ferranti. He also designed the huge, powerful generator engines that would run the station, making the whole project almost exclusively his work. Another key element of Ferranti's vision was for his station to supply large areas via the use of smaller substations. These were to be fed from Deptford, with power then being carried from each substation via mains cables. Miles of cabling was ordered during the construction of the power station, to be used to connect the various substations. Ferranti's vision called for higher voltages than had ever been previously transmitted however, and it soon became clear that the cables supplied were unable to take the strain. Again, Ferranti's solution was to take matters into his own hands, and he quickly designed a system of cabling that was suitable for such high voltage.

The station finally opened in 1891, but on a scale far smaller than Ferranti had planned. It was the result of an enquiry by the government-led Board of Trade. Chief among the board's concerns was the notion that huge areas of London were about to be supplied by just one power station. If it were to fail – a genuine concern considering the experimental nature of Ferranti's system – then thousands of consumers would be affected.

The board also foresaw that such huge dominance over London's power supply would leave little room for competition from other companies, creating a monopoly that would allow the London Electrical Supply Corporation to control electricity prices. Desperate to salvage their business, the LESC compromised by reducing the total output that Deptford would generate, and by halving the area of London it would supply electricity to. It was a move that angered Ferranti. He not only saw the Board of Trade's enquiry as damaging to his company, but also as detrimental to the development of the electricity industry as a whole.

Ferranti was dealt a further blow when it became clear just months after opening that Deptford Power Station was a huge commercial failure. Even at the agreed reduced capacity, the plant was far too powerful for the level of demand. The cost of generating massive amounts of power that could not be sold meant that the power station was constantly running at a loss.

Ferranti turbine engine at Deptford. (Image used courtesy of the Greenwich Heritage Centre)

With his vision for Deptford incomplete, and the public humiliation he was subjected to because of its failure, Ferranti saw no other option than to leave the company that he helped create. It continued to operate nevertheless, until its fortunes were finally reversed in 1925.

It was in this year that the LESC became one of the companies that were merged under the formation of the London Power Company (LPC). The acquisition of so many established companies gave the LPC control over many existing power stations, but they were also keen to make their mark with a brand new site of their own.

Despite the commercial failure of Ferranti's station, the site at Deptford was still seen as a prime location, and so the LPC began construction of a new power station which was to be named Deptford West (from this point on the original station was then referred to as Deptford East). Built on an even larger scale than its predecessor, the new site was the first giant power station of the modern era, with huge capacity that this time around would be used to its full potential.

It was designed by Leonard Pearce, whom the LPC made their engineer in chief in 1926. Born in Somerset, Pearce had a proven track record in the electrical supply industry. Having previously worked as engineer for various earlier electrical companies and London's underground railway, Pearce

Remains of the huge coaling jetty at Deptford.

designed a power station in Greater Manchester in 1920, showing for the first time that power stations could be designed with style in mind almost as much as function. When Deptford West opened in 1929 it was an architectural triumph, with a brickwork central structure that would later become standard, most famously at Battersea Power Station, of which Pearce was also involved in the design. The construction of the new building wasn't without incident however. Disaster struck when five workers were killed when a shaft being built close to the river collapsed.

Deptford West managed to succeed where the original station had failed to turn a profit, but the LPC retained Deptford East until finally decommissioning it in 1957. The newer building was expanded in 1948 to add increased capacity, and continued to operate until it was closed indefinitely in 1983.

The original building was demolished in the 1960s, followed some thirty years later by the destruction of the derelict Deptford West. Largely now forgotten, there are some remains that can be seen today, most notably parts of the huge coal ship jetty on the banks of the Thames. The cost of dismantling structures like this means that the river is littered with remains of this sort, and the sheer size of what remains today provides a great indication of just how grand the power station must have been. The London Power Company had its own small fleet of vessels serving Deptford, including one launched in 1932 named SS *Ferranti*, in honour of the great engineer.

The jetty can be seen by walking along the Thames Path close to Glaisher Street, but is perhaps best seen by taking a boat trip along the river itself. You can also find a brickwork electrical substation close by on Deptford Green, which probably also once formed part of the power station site. Outside London, a small section of the original Deptford East building has been preserved in Manchester at the Museum of Science and Industry.

The impact of what Ferranti achieved at Deptford has not been forgotten however, and his pioneering system for mass electricity supply is still in use today. He continued to be involved with the development of elec-

Traces of the former Deptford Power Station building.

tricity until his death in 1930, and a company he founded would later go on to manufacture what is regarded as the first commercial computer. His legacy was commemorated with the opening of Ferranti Park by Lewisham Council in 2005. It can be found near to the power station site, on Creekside.

Woolwich

The south-east London district of Woolwich is an area that has seen much redevelopment in recent years, with improved transport links, major investment and the success of London City Airport close by in Silvertown all helping to bring people and businesses back. But it is difficult for Woolwich to completely shake off its heavily industrialised history, and you don't have to look too hard to find derelict traces of its former past. The area once boasted several huge factories, a vast naval dockyard, the Royal Arsenal arms factory, and, of course, a power station.

The first steps towards bringing electricity to the district came in 1890 when a company known as the Woolwich District Electric Light Company Limited was granted permission to begin supplying power to light the streets of what was then referred to as the Parish of Woolwich. The firm chose a sight next to the Thames, close to where the Woolwich Free Ferry had recently opened in 1889. It is said that remnants of boat building and other artefacts from as far back as Roman London were discovered during the construction of the power plant, providing further evidence that Woolwich had been an area dominated by industry for centuries.

Coaling jetty at the former Woolwich Power Station.

Generating station at Plumstead.

The power station opened in 1893, and was later sold to the Metropolitan Borough of Woolwich in 1901 for a recorded sum of £80,000. The borough was founded a year before in 1900, and was keen to start selling power to its residents across Woolwich, Plumstead and parts of Eltham (today Woolwich is part of the Royal Borough of Greenwich, with parts of North Woolwich falling under the London Borough of Newham).

By 1920 the station was expanded to increase the capacity needed to meet the demands of the thousands of homes and businesses now wanting electrical lighting, and further expansion came in the next three decades.

Woolwich Power Station, however, wasn't the only electrical supply located in the area. In 1903 the local council also opened a somewhat groundbreaking electricity generating station in Plumstead. It also had a dual purpose as a rubbish incinerator, and it was the burning of the rubbish itself that generated the power needed to supply to the local vicinity (see also the section on Shoreditch Electric Light Station). It was later decommissioned, with Woolwich Power Station by this time well established and running at full capacity, although the Plumstead site was still used for rubbish incineration until the mid-1960s.

The station at Woolwich closed down in 1978 and was demolished in stages over the next two years. Parts of the site were kept intact, including offices and the main control room, but these also later disappeared, in 2002. The only thing that remains today is the coal jetty on the banks of the Thames, which can easily be seen near the foot tunnel.

The generating station at Plumstead is still intact and is now a Grade II listed building. It was used as council offices in the decades that followed its closure as an incinerator, but was later abandoned. At time of writing the building was being brought back into use as part of the Crossrail project. It can be seen on North Road, off White Hart Road.

Kingston

Kingston, in south-west London, is a town that suffered one of the most difficult transitional periods when London's districts were reorganised as part of the London Government Act of 1963. It is now part of the London Borough of Kingston upon Thames, but until 1965 the town was classed as being in Surrey, and was administered by the Municipal Borough of Kingston upon Thames. Also sometimes referred to as the Kingston Corporation, it was the local council who commissioned a power station to be built alongside the Thames in 1893. It was a modest affair with limited capacity, and as demand continued to grow plans were drawn up in the late 1930s to replace the original station with something more suitable. The need for a new facility was further highlighted

The former Kingston Power Station. (Image used courtesy of Neil Clifton)

in 1938 when most of the original site was destroyed by fire. The start of the Second World War did, however, halt progress in the building of its replacement. The new station, referred to as Kingston 'B', was commissioned by the council, although on opening in 1948 it was now under the control of the British Electrical Authority. It was opened with much fanfare, including an official ceremony attended by King George VI and the Duchess of York.

Despite having been replaced by Kingston B, the original site continued to be used until 1959. A decline in fortunes and the nationalisation of the industry also led to the eventual closure of the B station in 1980. The entire site lay abandoned for several years after, before finally being demolished in late 1994.

As is often the case when a disused building is demolished with explosives, it was an occasion that attracted hundreds of spectators. Amateur video footage taken on the day shows the station's two concrete chimneys fall dramatically one after the other, and the local newspaper is said to have covered the event with a special souvenir photo spread. It was perhaps a fitting end to a power station that had such a formal and regal opening decades earlier.

The station was located between what is now Canbury Gardens and the Kingston Railway Bridge. Nothing remains of the site, although it is possible that some of the boat mooring facilities close to Down Hall Road include traces of the former barge docks that were used to bring coal.

Croydon

Croydon Power Station is the collective name for two council-owned stations built by the Croydon Corporation. They were located on Purley Way, which was thriving at the beginning of the twentieth century as the centre of Croydon's heavy industry. A major gasworks and a steel mill were already established, and the area later became known for the manufacture of cars and aviation equipment. Purley Way is also home to Croydon Airport, which served as London's first international air travel hub after the First World War, only to be later eclipsed by Heathrow and Gatwick.

The first power station was opened in 1896. The Croydon Corporation was part of the County Borough of Croydon (later to become the London Borough of Croydon). They constructed the new plant as a way of supplying power to the public and the vast collection of factories and warehouses close by, having obtained the legal powers to generate electricity several years earlier in 1891.

The coal-fired generators were updated to increase capacity in 1924, but by the late 1930s plans were put in place to construct a second power station on the site. Progress was stopped at the outbreak of the Second World War, but by 1950 the new building was complete and officially opened a year later. It was given the name Croydon 'B' in order to distinguish it from the older building, which in turn was now referred to as Croydon 'A'.

Whereas A was designed with only function in mind, B followed a new trend emerging, where power stations started to be seen as buildings that should look just as powerful as the electricity they produced. In Croydon's case the design was the work of famed architect Robert Atkinson, whose Art Deco buildings had already made an impact across Britain.

The larger, more efficient Croydon B was to become the downfall of the older A station. Some of its cooling towers were demolished in the mid-1960s, an event that was watched by hundreds of

One of the remaining two chimneys at Croydon Power Station.

local residents. Its decline continued in the years that followed, and by the start of the '70s Croydon A was virtually unused. Closure finally came in 1973.

Croydon B continued to thrive for another decade before also being decommissioned in 1984. A few years of dereliction followed, until full demolition was ordered in the early '90s. An iconic part of the power station has, however, been preserved in the form of two brick chimneys that can be found close to what is now a branch of Ikea. Part of Atkinson's original design, the chimneys include elegant brickwork features that serve as a small reminder of what was once a building that would have dominated the local skyline. Regrettably the chimneys have been ruined somewhat by Ikea's corporate colours being placed around their tops. One additional reference to the area's history can be found in the name of a tram stop close by, appropriately christened Ampere Way.

A new gas-fired Croydon Power Station is now located nearby, opened in 1999. Similar to another current station at Brimsdown in Enfield, it is one of only a few active power stations remaining in the London area.

West Ham

West Ham was one of the most industrialised East End districts in Victorian London. The growing demand for electrical power in the area prompted a makeshift generator to be installed inside Canning Town Hall in 1895, which was used to power lighting for the hall itself, the local library and a small selection of businesses close by.

This basic but effective generator was the result of an application for permission and funding to provide power to the community, filed in 1892 by the County Borough of West Ham (often also referred to as West Ham Borough Council. Note that although today West Ham falls under the London Borough of Newham, this didn't happen until 1965).

By the end of the decade the town hall generator had become too small to meet the demand for electric lighting, which by now was illuminating streets, cloth makers' factories and other types of manufacturing plant. The erstwhile generator was expanded on several occasions but was never enough, and instead a new dedicated power station was opened in 1898, near to what today is the Olympic Stadium. It was located close to the still-standing Abbey Mills Pumping Station (see chapter 5). This also proved to be insufficient in meeting demand, however, and by 1899 the new station was already being expanded. Improvements included increasing power output with the installation of new coal-fired engines designed by Sebastian de Ferranti, similar to the ones at his pioneering power station in Deptford.

By the early part of the new century it was clear that the station at Abbey Mills would need another round of expansion if it was to meet demand. The only problem was that the surrounding area was now heavily built up, meaning that any expansion would require having to pay for houses to be cleared and public roads to be reconfigured. Instead, the borough council decided to close the site and replace it with a new dedicated station, to be known as West Ham Power Station.

Also sometimes referred to in archives as Canning Town Power Station, the new site opened in 1904 as one of the largest council-owned power stations in the UK. It was built alongside Bow Creek, close to where the River Lea meets the Thames. A common feature of most large power stations built in the early part of the twentieth century, the close proximity to water was crucial as it allowed for coal to be delivered by barge or collier ship. The site chosen had previously been a sewage farm, and some of the large tanks were retained as condensing ponds that worked in conjunction with two wooden cooling towers. Power itself was supplied by seventeen boilers used to work several generators of differing power, most of which were built by Ferranti's company.

By 1910, West Ham was supplying electricity to thousands of homes across Canning Town, Silvertown and Stratford. Yet their most lucrative contract was with the Port of London Authority, who purchased electricity to power lighting and machinery across the Port of London.

The demand for the huge amount of power that West Ham was capable of generating came not just from the major new thirst for lighting. The County Borough of West Ham had also founded West Ham Corporation Tramways in 1901, and needed the new power station to supply the current required to run its fleet of tramcars along more than 16 miles of track (note that despite its use as a power source for trams, West Ham is included here and not in the later section about power stations for transport, as the bulk of its output was non-tram related).

The station's success continued to grow until the start of the First World War in 1914, when street lighting restrictions in many public places led to a significant drop in the amount of electrical power being consumed. The station was instead used as part of the war effort, supplying power to factories and steel works converted to make ammunition and equipment.

The high demand of wartime use had taken its toll on West Ham Power Station, but the lifting of public restrictions after the war allowed the company to thrive once again. The continued success required that the ageing equipment be upgraded, and more generating power was also added in the 1930s.

The start of another world war would prove to have a far more devastating impact on the station however, as large parts were destroyed during the Blitz in 1940. The station was rebuilt after the war, yet as demand grew increasingly bigger it was decided that a second power station would be built on the same site.

Ownership had become the responsibility of the British Electricity Authority by 1949 as part of plans to nationalise Britain's electricity industry, and it was under their control that the new station opened in 1951.

Following a similar trend to other power stations – where new buildings were often opened alongside older ones – the original station was subsequently renamed West Ham 'A', and the new building West Ham 'B'. The A station was maintained, but its newer counterpart boasted modernised machinery and concrete cooling towers. It was soon supplying the bulk of the two stations' combined output. The B station also relied less on the use of the river, as coal was instead delivered by train.

The original station was later decommissioned, followed by closure of the newer site in 1983 – less than thirty-five years after it opened. The station did achieve one small claim to fame in 1965 though, when it was used as a location for a scene in *The Ipcress File* starring Michael Caine.

Both stations were demolished in 1984 along with all offices and other peripheral buildings. Today nothing remains, but the stations were located on what is now Electra Business Park, close to Stephenson Street and Bidder Street near Canning Town. The appearance of pylons, overhead wires and a small substation hint at a connection with the location's past.

Further along Stephenson Street is Star Lane Docklands Light Railway (DLR) station, which sits close to where West Ham Power Station's coal sidings once connected with the main line. All traces of this have long since disappeared.

Shoreditch Electric Light Station

The reputation of Shoreditch and Hoxton as being amongst the seediest areas of Victorian London have no doubt contributed to their ascendancy in recent years as being London's centres of cool chic.

Several buildings from the nineteenth century still exist today, built in an era when local residents were forced to live in impoverished streets alongside gasworks and other stink industries. One such building still surviving today is the Shoreditch Electric Light Station; a generator house built by the Shoreditch Vestry in 1896.

The Vestry was a parish administration that had held sway over the district since the seventeenth century. It would later be succeeded by the creation of the Metropolitan Borough of Shoreditch in 1900, itself later to be abolished and replaced by the London Borough of Hackney.

The station was built to provide street lighting for the local area, using a method of operation that was fairly ground-breaking for the time. The steam required to power the generating equipment was created by the incineration of public rubbish. It was, therefore, a pioneer in the process of energy-efficient

Above: Shoreditch Electric Light Station, now an events space.

Left: Shoreditch Electric Light Station.

recycling. It remains to be seen if this was the actual intention, or if creating steam from burning rubbish was simply far more cost-effective than having to purchase coal.

The facility was officially opened in 1897 by Lord Kelvin, who earlier in the century had designed efficient methods for measuring electric current. It ran successfully until 1940 when it was abandoned in favour of cheaper electrical supply, and sat derelict for some time after.

The building has since been fully restored and is now home to an arts and events venue known

Plaque commemorating the Light Station's history.

as Circus Space. The preserved exterior includes the original 'Vestry of St Leonard Shoreditch Electric Light Station' lettering above the entrance, and a motto in Latin that roughly translates as 'From dust comes light and power'. There is also a plaque that was erected by Hackney Council, commemorating its role in the history of London's electricity supply.

Acton Lane

The aesthetic quality of an abandoned industrial wasteland often makes buildings like disused power stations an attractive location for film and TV crews. Several of the stations and gasworks included here have appeared on screen, but their moment in the limelight is often nothing more than a blink-and-you'll-miss-it backdrop to a long forgotten scene.

However, in the case of Acton Lane Power Station, the building was used as the location of key scenes in two of the biggest films of the 1980s, and reference to them now seems to dominate most historical records of the site.

Its first appearance was in James Cameron's 1986 film *Aliens*, a sequel to the original classic of the late '70s. The power station was then used again in the 1989 blockbuster *Batman*, appearing as the chemical plant where Jack Nicholson's character falls into a vat of acid and emerges as the Joker.

Hollywood associations aside, the former station at Acton Lane was, in its prime, one of the most important power suppliers in west London. An original set of buildings were constructed by the Metropolitan Electric Supply Company who purchased a large plot of land on which to build the power station in 1897, located close to Harlesden. Rather confusingly, despite its

Partial remains of the former Acton Lane Power Station.

name having an obvious connection to Acton, the station was actually closer not only to Harlesden but also Willesden Junction. It is for this reason that it was sometimes referred to as Willesden Power Station. This, however, should not be confused with a different lost power station in Willesden, replaced in 1979 by Taylors Lane Power Station, which still exists today (see later section).

Acton Lane began operating in 1899 as a way of making up for the shortfall in capacity being suffered by the Metropolitan Electric Supply Company's existing smaller stations in central London. The company was established in 1887, and by the mid-1890s was supplying electricity via minor power stations in Soho, Bloomsbury and Marylebone. The firm desperately needed a more significant site however, and huge investment was pumped into Acton Lane to ensure it was powerful enough to meet the growing demand.

The new station enabled the company to supply to larger areas within central London, but also to extend its reach out westwards, so that it was now possible to cover places such as Acton, Hanwell, Greenford and Uxbridge.

Its location next to the Grand Union Canal allowed coal to be delivered directly, although the station also had its own railway sidings that facilitated the regular arrival of coal trains from mines in the north of England, via the Midland Railway.

In 1927 the Metropolitan Electric Supply Company became one of the firms acquired by the London Power Company, and Acton Lane was deemed to be of enough value to maintain. It was a decision in stark contrast to several other sites across London that were subsequently closed after the arrival of the LPC. The station was in fact expanded with a newer building, opened in 1950 and referred to as Acton Lane 'B'.

The entire site was decommissioned and closed down in the early 1980s. The various buildings sat derelict long enough to make their movie appearances, before being almost completely demolished a few years later.

An electrical substation does still exist at the location today, including some traces of the original site. The best way to see the current buildings is to walk along the Grand Union Canal towpath between Acton Lane and Old Oak Lane.

Hackney

The Metropolitan Borough of Hackney (today known as the London Borough of Hackney) was one of the most densely populated and deprived slum areas of London by the late nineteenth century. The arrival of the railways in 1840 had resulted in a huge swell of new inhabitants, creating cramped streets of housing for factory workers, built on what had been farmland only decades earlier.

Its position at the heart of industrial London did, however, mean that Hackney often found itself at the forefront of advancements in technology, in particular those related to the supply of power for lighting. Oil and gas had already been in use in the area for several years by the 1890s, providing lighting for factories and streets across Stoke Newington and what today is referred to as Hackney Central.

The Borough of Hackney was then granted the necessary powers in 1893 to generate its own electricity for public consumption, which led to the opening of Hackney Power Station a few years later in 1901.

The chosen location for the station to be built upon was a stretch of undeveloped land along the banks of the Lee Navigation (part of the River Lea), close to Homerton Hospital. The coal-fired facility was used primarily for supplying to Stoke Newington, but was later expanded to increase its capacity.

As with most lost power stations across London, nationalisation plans saw the running of Hackney Power Station transferred to the control of the British Electrical Authority in the 1950s, and a newer station was built in 1957 as a replacement for the ageing original, which finally closed in 1969. The new station, dubbed Hackney 'B', operated for less than twenty years before being decommissioned in 1976.

It is intriguing to note that Hackney Power Station was the subject of political controversy twice during its eventful lifespan. The Metropolitan Borough of Hackney voted Labour in the local election of 1919 – the first

Hackney Power Station before closure. (Photo used courtesy of Helmut Zozmann)

The remains of Hackney Power Station today.

to be held since the end of the First World War. But at the next set of local elections just three years later in 1922 a swing against the Labour party saw them lose all of their seats in the borough. It was a year in which Labour had gained great momentum in the run up to the general election, suggesting that the result in Hackney had bucked the trend. The Conservatives still managed to win the general election regardless, and in Hackney Labour had once again been beaten, this time by an alliance of reformists. It is perhaps no coincidence that Labour's election losses came in the same year that several of their local government officials had been accused of misusing revenue from the running of Hackney Power Station, which at this time was still owned by the borough.

The next wave of controversy came when the power station was forced to close for a period as a result of the 1972 miners' strike. It was one of several infamous disputes between the National Coal Board and the National Union of Mineworkers, and any stop in the production of coal meant that power stations were often unable to function. This in turn led to power shortages across the country, causing a national state of emergency to be declared until the strikes finally ended after almost two months.

Most of Hackney Power Station was demolished after closure, but a substation still exists on the site. This is located on the corner of Millfields Road and Mandeville Street. Along one side of the substation – facing towards a Hackney Council depot – it is easy to spot traces of the lost building, including a number of abandoned first-storey doorways, and six concrete stumps protruding out from the wall, each of which would have once helped support the former roof.

Wood Lane

The affluence in many parts of what is now the Royal Borough of Kensington and Chelsea meant that residents and businesses in the latter part of the nineteenth century demanded electrical lighting long before those living in areas of poverty. Two companies in particular worked fast to satisfy demand, and in 1888 both of them obtained the official permissions required to start generating electricity.

The first was the Kensington & Knightsbridge Electric Lighting Company, who built two small generating stations in 1890 to supply power to parts of the district. The other was known as the Notting Hill Electric Lighting Company. They constructed a modest station close to what is now Notting Hill Gate Tube station.

Within the space of just two years both companies needed to expand their operation in order to meet the ever-rising demand for power from their cus-

Painting of Wood Lane Power Station by artist Charles Holmes. (© Museum of London)

tomers. Neither had available space in which to enlarge their existing generating stations, so a radical decision was made in 1899 to merge and form a new, combined company known as the Kensington and Notting Hill Electric Light Company. The aim was to build a large new power station that would have enough capacity to allow them to dominate vast parts of west London, but first a suitable location had to be found.

In poorer areas of London, in particular the East End, residents often had little choice in being forced to live near dirty factories, gasworks or power stations. But this was west London, and the new company was acutely aware that building a large station in the wrong place would cause uproar.

A suitable plot of land was selected in Wood Lane, near to what today is the Westfield shopping centre (also see the later section on Wood Lane Generating Station). It was close enough for easy supply to the rich suburbs, but far enough away to ease their concerns.

Wood Lane Power Station opened in 1900 using an experimental type of cable for the transmission of current to various substations, based on the same principles as the wire first developed by Sebastian de Ferranti during the construction of his station at Deptford. Capacity was increased in 1906, and two years after that Wood Lane helped to supply power for the Franco-British Exhibition taking place nearby at what later became known as White City.

The absence of water in the vicinity meant that coal had to be delivered by train. At most power stations that relied on rail, coal was transferred from wagon to generator via cranes and conveyor belts. But the unusual layout at Wood Lane allowed trains to run directly through the building, dropping their shipment of coal along the way.

The Kensington and Notting Hill Electric Light Company was another of the firms absorbed by the formation of the London Power Company in 1925. The facility at Wood Lane didn't feature in the new company's long-term plans, and it was decommissioned in 1928.

The building was partially demolished in 1958. The remaining section was later leased to a number of different local businesses, including an electrical cable manufacturer. The entire complex was then demolished in 1979 and nothing now remains.

Wood Lane Power Station has however been immortalised in an oil painting by artist Charles Holmes, who lived close by in Ladbroke Grove.

Fulham

Fulham Power Station was a fine example of how a large industrial site can have such a dominating impact on a local community. In the fascinating book *Old Sands End, Fulham* (Sands End refers to the heavily industrialised area of Fulham where the power station and several factories were located), author Francis Czucha unearths a stunning collection of photos from the 1930s to the 1970s. They show everyday life in Sands End, plus special occasions like the street parties held for the Queen's coronation. The four chimneys of the power station can be seen in the background of many of the photos, acting as a dramatic backdrop to every occasion being documented.

Remains of the former buildings at Fulham Power Station.

As an employer of hundreds of people from the surrounding streets, it is obvious that a power station like Fulham helped to mould a community that was always aware of its domineering presence. When a building such as this disappeared, it was undoubtedly the end of an era, and an end to many of the traditional values that once defined British towns and cities. Today Sands End is much changed, but the legacy of the power station and many other industrial relics can still be found along its streets.

Fulham Power Station.

The station was council-built and operated, constructed in 1936 by Fulham Borough Council. It was built next to the Thames, located only a short distance from Lots Road Power Station at Chelsea Creek (the two stations were in fact often confused with each other. See the later section for more on Lots Road). It boasted four concrete chimneys in a straight line, which from a distance looked similar to a huge steam liner being moored on the river.

It wasn't the first power station to be built on this site, however, as the council had previously opened a smaller facility here in 1901. The Metropolitan Borough of Fulham (superseded in 1965 by the London Borough of Hammersmith and Fulham) has its roots in the Fulham Vestry, one of several civil parishes that had administrative powers over many parts of London. The vestry obtained the necessary permission in 1897 to supply electricity for lighting, becoming one of the earliest local authorities in London to do so. By 1915 the station was running at maximum capacity, supplying to large parts of Fulham, with agreements also in place for the provision of power across the river to Battersea when needed.

Greater capacity was soon required as the area continued to develop. The council responded by opening the new station as a replacement to the original. It contained sixteen coal-fired boilers capable of generating huge amounts of power, so much power, in fact, that it became the biggest council-owned power plant in the country.

In addition to its main use as an electricity supply, accounts from employees also suggest that the council was able to earn extra revenue by allowing the power station to be used as an incinerator for the burning of old British and US banknotes taken out of circulation. Staff are also said to have performed the more sinister task of looking out for suicide attempts along Wandsworth Bridge, alerting the police each time they spotted a potential victim.

The station sustained bomb damage during the Second World War, but was rebuilt and ran successfully for several years, before being decommissioned in 1978 as part of nationalisation plans for the industry. Most of the site was demolished in the 1980s, including the four iconic chimneys.

It brought an end to electricity supply in Fulham, but the demolition process also resulted in a scandal that would ruin its legacy. In a story that was similar to hundreds of former miners across the country, the relatives of several retired workers at the power station claimed that working conditions at Fulham (and no doubt at other power stations) had contributed to the early death of their loved ones from lung cancer. It was a theory given substantial weight when large amounts of asbestos were removed from the site during demolition, leading to complaints by local residents about heavy dust clouds caused by the works.

The debate made it to the House of Commons in 1983, where a series of errors in the monitoring, removal and safe disposal of the asbestos were uncovered. The controversy resulted in more stringent rules being introduced regarding the decommission of old power stations, yet it did little to help the health issues being faced by some of Fulham Power Station's former workforce.

Today the chimneys may be long gone but part of the power station still remains intact, including the front of the building, which can easily be seen on Townmead Road. Complete with 'Fulham Borough Council Electricity Department' lettering along the top, the facade now hides a small substation that serves as the only significant trace that a major power station once stood here. The rest of the building was located on what is now a branch of Sainsbury's and the Big Yellow Self Storage facility.

It is also possible to walk along the river behind Sainsbury's. It is now a stylish stretch of waterside development that includes the exclusive Harbour Club, Chelsea, but it was here where the power station's many cranes lifted tons of coal every day. Unlike most other power stations, which often had to use collier ships owned by coal companies, Fulham Power Station had its own fleet of ten ships, each painted in the same colour with the words 'Fulham Borough Council' on the side. Two of the colliers were damaged during the Second World War, causing the death of a crew member in 1945. The ships were either scrapped or sold on after the power station closed, and it is unlikely that any of them are still in existence today.

The wooden coaling jetty from which the colliers would have unloaded still remains today, and is easy to spot behind Sainsbury's. There is also a plaque

Above: Former coaling jetty at Fulham.

Left: Plaque highlighting the former power station site.

on the side of the supermarket building that commemorates the site's former use as a power station.

It is important to note that the location of Fulham Power Station is often described as being a large derelict and graffiti-covered building that lies further along Townmead Road. Its resemblance to the remains of Lots Road Power Station mean it is an easy mistake to make, but this building is in fact a former food packing factory and brewery. Even further along from here can be found a series of fuel storage tanks, best viewed from Wandsworth Bridge. This is all that now remains of a former Shell Mex storage depot that was in operation at the same time as Fulham Power Station, contributing to what was no doubt a highly polluted stretch of the river.

Blackwall Point

Blackwall Point was a power station located on the Greenwich Peninsula, within walking distance of what today is the O₂ Arena and North Greenwich Tube station. It is an area whose regeneration hides decades of industry, with a wide range of factories and wharves, a major gasworks and the power station, all once situated on the site of what has become one of London's most popular attractions.

The power station was built in the 1890s by the Blackheath and Greenwich Electric Lighting Company, and finally opened in 1900. It was a minor company, set up solely to supply power to the areas included in its name, and was later one of several firms absorbed by the Metropolitan Electric Supply Company (formerly the South Metropolitan Electric Supply Company).

The original building was closed down in 1947 and replaced with a newly built power station located on the same site, constructed in 1952 by the British Electrical Authority. Taking the lead from successful experiments at US power stations, the new site at Blackwall Point developed the concept of using pulverised coal to fuel its boilers. Prior to the introduction of this new process, coal would have been fed into the boilers used to power the generators in medium-sized lumps. While effective, large portions of each coal shipment would be wasted by remaining unburned, due to the size of the lumps.

The new process involved the coal being crushed into powder, which made it easier to burn, with less wastage. Special milling equipment was installed at Blackwall Point to crush coal after it had been delivered by collier, before then being taken to the boiler room. It was the first power station in London to trial the new technique, and indeed one of the first in the world. The use

Abandoned coaling jetty at Blackwall Point.

of pulverised coal has since become the standard for many power stations around the globe.

The station closed down in 1981 as part of the nationalisation of the industry. It was demolished a few years after closure, with no trace left of any of its buildings. A significant coal ship jetty does still exist though, and this can easily be seen by walking along the Thames Path, close to the London Soccerdome and the Greenwich Peninsula Emirates Air Line station. Eagle-eyed viewers will also spot the coaling jetty in the famous opening credits of *EastEnders*.

The location of the former Blackwall Point Power Station is almost directly opposite to the site of the demolished Brunswick Wharf Power Station, on the other side of the Thames. Taking into consideration how close these two sites are to the old power stations at Deptford and Greenwich, it becomes clear just how industrial this stretch of the river once was.

Stepney

Stepney Power Station was located in the east London district of Limehouse, an area known for its centuries old association with the Thames. It was home to hundreds of shipbuilders and rope makers in the eighteenth century, and the opening of Limehouse Basin in the nineteenth century provided a link between the city's growing canal network and the river. It brought new industry and a sharp rise in population.

This lead to an inevitable demand for electricity, prompting the Metropolitan Borough of Stepney to respond by filing an application to start generating power. Construction of a dedicated power station started in 1907 and was quickly completed in the same year, with the generators turned on soon afterwards. The station was used to supply local homes and businesses across the borough, with extra revenue also earned from selling additional power to the Metropolitan Borough of Bethnal Green.

The local council in Bethnal Green had obtained powers to generate themselves in 1899, but never actually built a power station of their own. It was therefore decided instead to simply buy electricity from Stepney, although it seems doubtful that this would have been deemed a viable solution in the long term.

The new power station brought thick smoke and smog to an area already choking from gasworks and factories long established close by. It caused major concern for the health of local residents, resulting in the council having to adapt the power station so that it had only one chimney.

This solitary chimney did at least allow Stepney to stand out from every other power station along this part of the Thames, providing a useful navigational marker for boats and barges moving from river to canal. In fact, the station's location next to both water routes allowed for coal to either be delivered by collier or barge, and small traces of the coaling jetty may still be seen today, close to Limehouse Basin.

Like other power stations in London, nationalisation saw ownership of Stepney Power Station taken away from the local council, and it was closed down in 1972. It was demolished several years later as part of a huge redevelopment scheme that has transformed Limehouse Basin into a vibrant marina, overlooked by expensive waterside apartments.

Virtually nothing now remains to suggest that this was once a hive of industry, although you don't have to venture too far along the canal until the stylish apartments begin to be replaced by old abandoned factories.

Note that in some contemporary records, Stepney Power Station is referred to as Limehouse Power Station.

Brimsdown

Although now part of the London Borough of Enfield, the north London town of Enfield was classed as being part of Middlesex at the beginning of the twentieth century. Its close proximity to the canals of the River Lea meant that Enfield grew as an industrial town in the late nineteenth century, with brickworks and a large arms factory among the biggest employers of local people.

A factory close by in Ponders End even played its own key role in the development of electricity supply by manufacturing light bulbs. Owned by Edison Swan – a company with ties to electricity pioneers Thomas Edison and Joseph Swan – the factory was also later adapted for the manufacture of valves for use in radio, TV and radar.

The closeness to a water source made Enfield the ideal place for a power station, and in 1904 the nearby area of Brimsdown was chosen as a suitable location for a new plant being built by the North Metropolitan Electric Power Supply Company. The firm had been founded by an Act of Parliament four years earlier in 1900, and later was to become one of the most significant electrical power companies operating in and around London.

Crucially, their success came from focusing on supplying power to districts on the outskirts of London, rather than in the city itself. As a result, when the company finally opened Brimsdown Power Station in 1907 it was in order to supply power for electric lighting across parts of Middlesex, Hertfordshire and Essex. Brimsdown also provided Enfield town with its first taste of street lighting, which in turn encouraged more industry to the area.

The station was of a fairly modest size on opening, with coal-fired generators supplied by fuel delivered along the Lea canals via barges. In addition to selling its power for lighting, the station was also used to provide traction for the North Metropolitan Tramways Company, whose ownership had ties with

Brimsdown Power Station. (Image supplied by The Enfield Society)

that of the power company itself. Having already successfully run a steam-powered tramway route from Wood Green to Ponders End, the company was now ready to electrify the route, and therefore the new power station proved to be the perfect solution (similarly to West Ham Power Station, Brimsdown is included here and not in the transport section as it was primarily a station used for public supply).

The power station was later extended to boost capacity, but it was then replaced with a new building in 1932. The move towards nationalisation in the late 1940s and early 1950s saw the ownership of Brimsdown pass to various different organisations. By the mid-1970s it had also become something of an industrial relic, with tired machinery no longer able to operate at a rate efficient enough to be profitable.

It was decommissioned and closed permanently in 1974, with demolition coming soon after. The site was later redeveloped and nothing now remains of the original station. A new power station was built, however, in 1999 on part of the original site, known as Enfield Power Station. The newer station is gas-fired, and is one of only a small group of large scale power stations still in existence in the London area.

Barking

Barking has been a heavily industrialised area of London since the latter part of the nineteenth century, when chemical manufacture and other stink industries replaced the fisheries and boat makers that had operated in the vicinity since as far back as the fourteenth century.

The area's new-found dominance as an industrial hub made it an obvious choice for a power station, and its origins in London's electrical supply history began with a small generating station constructed in 1897 on the banks of the Thames at Creekmouth. It was a council-owned plant built by Barking Urban District Council (later to become part of the London Borough of Barking and Dagenham), in order to supply lighting for homes and factories in west Essex.

The station was later one of many absorbed by the County of London Electric Supply Company. Founded in 1891, the firm took ownership of various smaller companies across London in the early part of the twentieth century. By the 1920s it had obtained the necessary powers to expand their reach towards the fringes of east London and beyond.

Eager to capitalise on the prime location of the original council-owned power station, the new owners opened a modern, larger station close by in 1925. It began operation with little fanfare at first, but a few months later it played host to a grand opening ceremony said to have been attended by no less a dignitary than King George V himself. The new plant allowed the company to sell power to large parts of Essex and Kent, and it wasn't long until new capacity was required. The solution was to build another new power station on the site. This opened in 1939, and became known as Barking 'B'. The 1925 station, by now referred to as Barking 'A', was rendered virtually obsolete, as was the original council station.

The collective power station buildings at Barking were to play a crucial role in protecting the area during the Second World War, albeit not perhaps an intentional one. With coal delivery becoming increasingly difficult during the war years, several power stations were forced to run at low capacity. Barking, however, was able to take advantage of coal that had been stockpiled in preparation for war, stored close by at Dagenham Docks.

It was able, therefore, to operate at full capacity, and it is said that the smoke from its chimneys helped to obscure the view from above of several key potential targets for German bombers. The most significant factory in the vicinity of the power station was the massive Ford motor works in Dagenham, which was being used during the fighting to manufacture thousands of military vehicles. Thanks to Barking it managed to avoid bomb damage, as did various other local factories.

Further expansion led to a third power station being built in 1954, by which point the entire operation had been absorbed by the British Electrical

Derelict remains of Barking Power Station.

Authority (BEA). Following the rather mundane but practical naming convention of its predecessors, the third station was named Barking 'C'.

The huge size of the site at Barking meant that it was a significant source of employment for people in Dagenham, East Ham and other local areas. Hundreds of workers were to lose their jobs, however, when Barking Power Station was forced to scale back in the face of a steep decline in coal-fired electricity supply. Buildings A and B finally closed in 1976, followed a few years later by C in 1981.

It dealt a huge blow for local employment. Large-scale power stations, gasworks, factories and other industrial plants were crucial to the livelihood of thousands of Londoners, in an era where men were almost exclusively the sole breadwinners in most family units. In fact, Barking Power Station was one of several employers at the time with a policy of only employing men until the 1940s. The belief was that electrical power generation was an industry too dirty for women, and it was a situation that meant even general office and secretarial roles at Barking were filled by men.

Major employers of this size also provided workers with the opportunity to join sports teams and clubs, therefore the decline of a power station as large as Barking would have had a deeper social impact than simply unemployment.

Power generation returned to the area in 1995 with the opening of a new gas-powered station known as Barking Reach, located close to the original

A, B and C site. Almost all of the disused Barking Power Station buildings had been demolished in the 1980s. There are some surviving parts, however, including one of the former switch houses, various office buildings and a control room complete with some equipment still intact. These can be seen along River Road, next to the location of Dagenham Sunday market.

Battersea

Battersea is not only the most famous disused power station in London, but also one of its best-known buildings outright. Its distinctive chimneys are recognised by people around the world, and they are used as a visual signifier of the city just as frequently as the London Eye, the Gherkin, St Paul's Cathedral, Tower Bridge and the Shard.

Its history began with the formation of the London Power Company (LPC) in 1925. The company soon established itself as the industry leader by building a power station of its own at Deptford (see the Deptford Power Station section), which opened in 1929. But two years prior to this, a proposal for another new station was also approved, and construction work began in earnest soon after. The station at Deptford was huge, but it was a project tainted by having been built alongside an earlier power station that had been deemed a failure. The new station therefore called for a new location, big enough for a mega-sized power station that would be capable of generating electricity on an epic scale.

The chosen site was a strip of derelict industrial wasteland next to the Thames, in the Nine Elms area of Battersea in south London, at the time part of the Metropolitan Borough of Battersea. It was an area that had been home to heavy industry since the sixteenth century, when mills and breweries began to dominate Battersea and nearby Wandsworth. The Industrial Revolution later brought candle, clothing and ceramics factories to the district. By the nineteenth century the area was also dominated by London's growing railway, acting as a major junction point for train lines running from the south to Waterloo or across the river to Victoria.

Nine Elms rapidly became one of the most polluted parts of London, with a railway goods yard and a large scale gasworks (see later chapter) among the biggest contributors. The prospect of a giant power station being added to the industrial landscape caused immediate concern, but the site was purchased nonetheless. It was close to the gasworks, on a piece of land once occupied by the Southwark and Vauxhall Waterworks Company, who supplied water to south London before moving to a new site in Hampton.

The concern over pollution – a surprisingly progressive and modern debate considering the time – dealt the London Power Company much negative

The iconic towers of Battersea Power Station.

press coverage, and it wasn't the only issue being raised. The LPC argued that Battersea was an appropriate place to build a power station because it was close enough to central London to best supply to the consumers within it. But those in protest claimed that its location so near to the centre would ruin the view of historical London, and be in stark contrast to other famous landmarks. As with most large, modern buildings constructed in London over the last century, the effect that the new power station might have on views of St Paul's Cathedral was of particular concern, with the Archbishop of Canterbury chief among the protestors.

The LPC's response was to silence the critics with some positive PR spin that aimed to reassure the public on every issue raised. The environmental concerns were eased with assurances that Battersea Power Station would utilise a ground-breaking new method known as 'scrubbing', where sulphuric emissions would be treated and broken down into a less hazardous discharge that could be released safely into the Thames, instead of into London's atmosphere. Ironically, tests later revealed that the effect this process had on the river was actually far more damaging, but in the late 1920s it was a strong enough argument to win over protestors.

Next came the task of convincing London that the new power station would complement the skyline rather than destroy it. The solution was to employ the services of architect Giles Gilbert Scott. Already a designer of some acclaim, the LPC's plan was to show that with Scott's involvement, the public could rest assured that the new power station would be designed with style in mind.

They also knew that Scott's profile was riding high at the time, thanks to the recent introduction in London of his K2 red telephone box, which would soon become an icon of the city and, indeed, country. Lastly, Scott's previous work had included either the full or partial design of churches in Harrow, Kensington, Gospel Oak, Chiswick and other London districts, and further afield in Yorkshire, Norfolk and Somerset. His work on Liverpool Cathedral was already impressing critics, and it was this religious connection that was used to ease the fears of the Archbishop.

The power station was constructed in two phases: Battersea 'A' and Battersea 'B'. Although it was decided from the very start to build them separately over a period of several years, the finished article would work together as one building of two identical parts.

Construction of Battersea A started in 1929, and by 1933 it was clear that Scott had created a masterpiece. His design drew comparisons to his work on religious buildings, leading critics to describe Battersea Power Station as a 'cathedral of bricks' and a 'temple of power'. Even the inclusion of a chimney at either end of the main brickwork building was seen as a feature comparable to a set of church spires.

Whilst Scott was collecting accolades for his design of the power station's shell, the interior design was also attracting the attention of critics. It was the work of a team of designers headed by the LPC's Chief Engineer Leonard Pearce, who also designed the new station at Deptford. Filled with ornate detail and Art Deco flourishes, Pearce's work at Battersea further underlined that this was to be a building worthy of taking its place among London's long line of historic structures.

Completion of Battersea A had been a long and dangerous process that had claimed the life of several workers, but the station finally began generating in 1933, and was running at full capacity two years later.

Construction of Battersea B was delayed by the outbreak of the Second World War, but work finally began in 1945. Post-war austerity measures meant that the interior was built on a far less lavish scale as the first building, but from the outside the overall structure now boasted two identical brickwork sections, four chimneys and a vast turbine hall connecting them in the middle.

When Battersea B began supplying in 1955, the combined power station was one of the largest and most powerful sites in the UK, although by this time nationalisation plans meant that the London Power Company had been replaced by the British Electrical Association.

One of the elegant and preserved control rooms at Battersea.

Battersea Power Station control room 'B'.

By the mid-1970s the massive boilers at Battersea were coming to the end of their lifespan. The rapid decline of coal-fired power stations also meant that the station was running at low capacity, and the cost of replacing the boilers and other maintenance work became impossible to justify. Battersea A was officially decommissioned in 1976, followed by Battersea B seven years later in 1983.

It was an era when the importance of preserving historical buildings from the nineteenth century onwards had not yet been fully appreciated, and plans were quickly put into place to demolish the power station. It was a controversial proposal that managed to cause even more protest than when Battersea was being built. Its future was eventually secured, however, when the site was awarded a Grade II listing.

It had been in the latter part of its working life that the building started to gain its iconic status, thanks largely to a string of appearances in music, film and TV that continues to this day. Power outages caused by a fire at the site even managed to delay the launch of BBC 2 in 1964, with Television Centre across town in White City suffering a huge blackout. The most famous appearance of Battersea in pop culture came in 1976 when it was featured on the cover of the Pink Floyd album *Animals.* A concept album that reflected on George Orwell's *Animal Farm,* the cover depicts a huge inflatable pig floating between two of the chimneys.

Today Battersea Power Station stands almost completely intact, and its four chimneys are now an integral part of London's skyline. On the outside the vast brickwork shell has aged fairly well, with only superficial damage and the obligatory smashed windows that all abandoned buildings seem prone to. The building is unfortunately in a far sorrier state when you step inside. One of the many failed attempts at redeveloping Battersea came with a bid in 1989 to turn the site into a theme park. The deal fell through, but only after construction work had begun, which included the complete removal of the central roof. It was never replaced, which means much of the interior of the power station has been exposed to the elements ever since. Decades worth of rain, cold winters and occasional sun have taken their toll, rendering parts of the building a damp, overgrown mess that has led to it being included on several endangered historical building lists.

There are, however, parts of the building still protected and well preserved inside, including two control rooms with hundreds of switches, levers, dials and gauges still in situ. Several London Power Company signs and plaques also still adorn the walls, and outside, next to the river, Battersea's two large coaling cranes still stand proud.

The power station's immense size means that it can be seen for miles around, but it is best viewed in all its glory from Grosvenor Road in Pimlico (on the opposite side of the road to the Western Pumping Station – see

Above: Former coaling cranes outside Battersea Power Station.

chapter 5). From here it is possible to peer through the railings and across the river to see this spectacular building from its most recognisable angle. It is also possible from here to look at the power station in context with the adjacent gasometers of the former gasworks at Nine Elms (covered later).

Walking along Chelsea Bridge also provides a fantastic view. From here, with boats gliding along the Thames below, and trains running along Grosvenor Railway Bridge between here and the power station, it doesn't take much imagination to picture how the scene would have looked during Battersea Power Station's heyday: the chimneys billowing smoke, colliers on the water delivering coal, steam engines taking passengers into Victoria station, and diesel engines moving freight to and from Battersea Wharf Goods Depot that once stood close by.

In fact, taking a train ride between Victoria and Battersea Park station is another great way to get a close-up view of the power station. It is also possible to see the other side of the building by walking around the industrial estate on Kirtling Street and Cringle Street, both of which can be found along Battersea Park Road. Regrettably, tight security around the power station site means it isn't currently possible to walk along the river directly in front of the building. The site is, however, sometimes used as a corporate and sporting venue, perhaps worth the often high ticket prices simply to get up close and personal with London's most famous derelict building.

The botched theme park bid and various other failed attempts to redevelop the power station site are looked at in chapter 7.

Bankside

The Tate Modern helped bring art and culture to the masses when it opened in 2000, housed inside a huge brick building that was formerly Bankside Power Station. The journey from power supply to the world's most popular art gallery is one filled with twists and turns, and one of the most successful examples of urban regeneration.

Bankside is the name given to part of the south bank of the Thames, commonly accepted as stretching from Blackfriars Bridge to London Bridge. It was a dirty, heavily polluted industrial landscape in the late nineteenth century, made worse when a power station was first built here in 1892. Located on land that is now a series of apartment blocks behind the Tate Modern, the station was built by the City of London Electric Lighting Company. The firm had been created specifically to supply power to the City of London on the opposite side of the river.

As the financial capital of the world, the City was one of the most lucrative areas for early electrical supply companies. Choosing to locate the power station on Bankside was a decision that suited everyone. It was close enough for the company to supply its wealthy City customers with little effort, but far enough away for those same customers not to complain about the smoke, smell and noise that having a power station on their doorstep would have created, had it been built in the City itself.

The only people the decision didn't benefit were the poor residents of Southwark that lived close to Bankside, and the arrival of the power station would see them suffer the effects of poor air quality for the next fifty years. By the mid-1940s the station at Bankside had been extended several times in order to increase its capacity. It was an immense operation with no fewer than eighteen concrete chimneys, each billowing smoke as the coal-fired boilers worked ever harder to meet demand. Much of the other industry that once surrounded the station had by now disappeared or been destroyed in bombing raids during the Second World War. In fact, almost the entire Bankside area had become a desolate wasteland, bringing yet more misery for the local residents.

The destruction caused by the war was looked upon by local governments and town planners as a chance to start again with a clean canvas. Several towns and cities across Britain were rebuilt according to ambitious and often controversial new designs, including the construction of thousands of tower blocks in major cities. It was a difficult process that split opinion, creating a wide gap between those with wealth and those without. While some agreed

Right: Bankside
Power Station during
construction.
(© *Daily Herald*
Archive/Science &
Society Picture Library)

Below: The Tate
Modern, previously
Bankside Power Station.

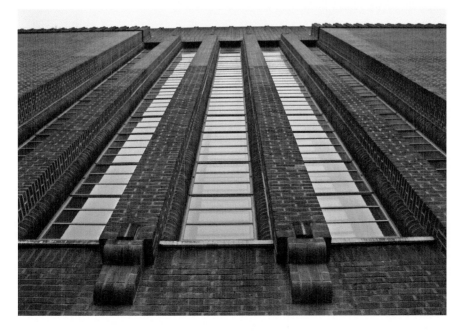

that post-war Britain should be reconstructed as something considered better than before, there were many who simply wanted to go back to what their life was like before the conflict.

But in an area such as Bankside it would have been hard for even local residents to deny that their quality of life would be improved by major regeneration. Plans were drawn up for derelict and abandoned factories to be cleared, and for a series of new cultural centres to be built along the river. The latter buildings would later open further along the river closer to central London, in the form of the Royal Festival Hall and the rest of the Southbank Centre. But for now, Bankside was at last being cleaned up.

The redevelopment plans left Bankside Power Station as a huge white elephant that no one was quite sure what to do with. It hardly fit the profile of the new, less polluted and more enjoyable Bankside being planned, but electrical power was of course still needed. More than ever in fact, as predictions showed that London would demand higher levels of electrical power than ever before once post-war austerity measures were removed. The controversial decision was therefore made that not only would the new Bankside area keep its power station, but it would also be rebuilt to a larger and more powerful scale.

Outrage and protest soon followed, with the same concerns as those voiced a decade earlier during the construction of Battersea Power Station. If London's cultural and religious leaders had been opposed to a huge building obscuring St Paul's Cathedral despite being several miles away from it, then they must have been horrified by the prospect of a similar structure now being proposed directly opposite. A petition was circulated, and it wasn't long before hundreds of influential figures among London's elite had voiced their opinion.

The defining moment in the future of Bankside came in January 1947 when the country was in the midst of an energy shortage. Production at coal mines had slowed down since the end of the war, leading to a shortfall in coal delivery to homes and businesses. The timing couldn't have been worse, as the entire country was suffering from the effects of a cold snap that caused a huge spike in demand for power, as people were forced to use electrical heating devices in the absence of coal. Considering power stations also needed coal to operate, most were unable to meet the extra demand, including those in London. The conclusion was simple. Bankside was needed desperately, and it had to be a station that ran on oil instead of coal. Approval needed to start the project was fast-tracked as the crisis got worse, and construction began later in the year.

Similar to Battersea, assurances were made that the new station, officially named Bankside 'B', would be designed in a style that would fit its surroundings. The services of Giles Gilbert Scott were called upon once again, with a brief that he should make the exterior something truly special. The result was a building that took the concept of power stations being cathedrals of power

Tate Modern at the former Bankside Power Station.

to a new level, with more than four million bricks, and windows reminiscent of those often found behind the altar in a church.

In the centre was a huge rectangular chimney that would prove to be the crucial element in helping to convince the protestors. The original plan was for Bankside to have column-shaped chimneys at either end, typical of most other power stations. Realising that one chimney would be far less obtrusive; Scott revised the design to include a single smoke stack, with a rectangular shape that blended well with the rest of the structure. As a further sign of sensitivity towards St Pauls, the chimney was also topped off at a lower height. The new power station, being oil fired, also assisted in the design process, as there was no need for a coaling jetty to be constructed on the river bank, and no demand for a dirty coal shed. Lastly, the building was set back from the river, to allow for other buildings to be constructed in front that could perhaps hide the power station if needed.

Bankside B opened in 1952 while still unfinished. The original station building, now referred to as Bankside 'A', was retained in service during construction, to ensure no break in supply. The older station was then decommissioned in 1959 and demolished soon after. Work on Bankside B was finally completed in 1963, but within twenty years the power station was decommissioned as a result of high prices for oil.

The building faced demolition as it stood in the way of redevelopment. But in a dramatic change of heart from just three decades earlier, London now embraced Bankside B as a historical icon that needed to be preserved, and various campaigns were launched to save the building. It was too young a structure to be eligible for listed status, however, and its fate hung in the balance for several years as dereliction set in.

Salvation came in 1994 when the power station was purchased by the Tate Gallery as a home for their new showcase of contemporary art. Work was complete by 2000, and the gallery was opened with a high-profile ceremony by Queen Elizabeth II. Tate Modern has since become a huge success with millions of visitors, rejuvenating an area in the very building that many feared would destroy it.

Fully aware that the building itself is a piece of art, the owners of the Tate Modern have left Bankside B largely untouched, and in fact a small part of the building still houses an electricity substation. An original crane has been preserved inside, and is sometimes used for lifting large installations on display inside the gallery's showpiece; the great turbine hall, which can be visited for free.

The Millennium Bridge was built next to Tate Modern in 2000, providing a direct link across the river to St Paul's. It is a fitting tribute to Giles Gilbert Scott's concept that Bankside Power Station should complement Christopher Wren's masterpiece, rather than obscure it.

Brunswick Wharf

Built in the same spectacular brickwork style as famous London power stations like Battersea and Bankside, Brunswick Wharf opened in 1952 after almost five years of construction work. It was located in Blackwall on the Isle of Dogs, on almost the exact spot from where three ships set sail in 1606 to discover the new world, establishing the first English colonies in America the following year. The area later became the East India Docks, which were already in rapid decline when the power station arrived.

It was a council-run operation built by the Metropolitan Borough of Poplar (later absorbed by the London Borough of Tower Hamlets). But in contrast to council power stations such as the ones at West Ham and Fulham, whose ownership only later became the responsibility of the British Electricity Authority (BEA) as the industry moved towards nationalisation, Brunswick was opened by the BEA from the beginning.

By the mid-1950s the station was supplying power to local homes in the area, and to various factories in and around the docks. Drawing further comparisons to Battersea, and also Lots Road, Brunswick Wharf boasted two concrete chimneys that could be seen from miles around. It was therefore a key

Possible remains of the coaling jetty at Brunswick Wharf Power Station.

marker point for anyone navigating their way along the Thames. Its location on the banks of the river also allowed enough space for easy delivery of the incredibly large shipments of coal needed to run the plant, especially after the station was upgraded to greater capacity in 1957.

Many power stations tended to be connected to the canal system, in particular those elsewhere in east London. The man-made and narrow construction of the canals meant that these stations had to rely on barges alone for coal delivery. Brunswick Wharf and other stations along the Thames were able instead to accommodate huge collier ships, built or converted specifically for bringing coal from the north, and Newcastle in particular. The Brunswick station included a vast concrete wharf equipped with three large cranes.

The power station's demise was ironically the result of ambitious plans to increase its capacity. Having been coal-fired from the beginning, new owners the Central Electricity Generating Board (covered in more detail later) decided to adapt the plant in 1970 to become oil powered. The Middle East oil crisis three years later meant however that oil prices surged, rendering the newly converted fuel far from economical.

The entire station was closed in 1984, and subsequently demolished for redevelopment as part of the gentrification of the Docklands.

Nothing at all remains, although it is possible that the disused jetty standing in the river close to the entrance of the old East India Dock Basin was part of the site. Considering its similarity to the design of Battersea Power Station, which has become a genuine London icon, it is somewhat surprising that such a prominent building has disappeared so conclusively. The presumption is that it didn't feature in the grand redevelopment plan for the Blackwall area and the Docklands overall, and it is unclear if simply converting the building to apartments was ever considered as an alternative to demolition.

A connection to the site's electrical history has, however, been preserved in the names of three modern apartment buildings close by: Elektron Tower, Proton Tower and Neutron Tower. Near to the river jetty is a residential complex known as Virginia Quay, in front of which can be found a monument dedicated to the new world explorers, and several surrounding streets built on what used to be the power station have also been named after those who helped found the early American colonies. Virginia Quay itself is also a nod to the USA connection, as indeed is the naming of the power station after Brunswick.

Bulls Bridge

A relative newcomer compared to other lost power stations in London, Bulls Bridge Power Station had a short and rather uneventful life that lasted a mere twenty years. Sometimes referred to as Hayes Power Station, due to its location in the Hayes area of west London, the power station was built in the late 1970s by the Central Electricity Generating Board (CEGB), which by now had taken control over the British electrical supply industry.

The company constructed the station on a large piece of land purchased from Middlesex County Council, alongside the Grand Union Canal. The power station was named after Bulls Bridge itself, a white painted bridge over the canal close to where the station was located. The bridge was at one time one of the most important junctions on the canal, where commercial canal boats would congregate on their way along the huge network of waterways that ran between London and Birmingham.

The station was one of the first gas-powered sites in London, paving the way for current stations fuelled by gas at Barking, Croydon, Enfield and Willesden. It was perhaps a little too ahead of its time in the 1970s, however, and it was closed down less than five years after opening.

The move towards gas power may have been slightly premature, but it was to become more widely used a few years after this station's closure. Coal-fired power stations were in rapid decline by the early 1980s, but London and many other cities still depended on them as their only source of electrical power. But stations that ran on coal were at the mercy of miner's strikes, which could quickly lead to a scenario where no coal could be delivered, and therefore no electricity could be produced.

In one of the most violent clashes between the National Coal Board and the National Union of Mineworkers, workers from several mines across Britain had again taken strike action in 1984. It wasn't long before the lack of fuel was felt in the nation's capital. A temporary solution was to reopen Bulls Bridge as a way of accounting for some of the shortfall in capacity. Its modest size meant it was hardly in a position to save the day on its own, but it certainly helped, and further highlighted calls for remaining coal-fired power stations to be converted to either gas or oil.

Later, privatisation of the industry saw ownership of Bulls Bridge transferred to Powergen, who maintained the station until decommissioning it in 1993. The entire plant was demolished within a few years of closure and is now home to an industrial estate. Nothing remains, but the view of both sides of the canal from the road bridge on North Hyde Gardens indicates where the power station once sat. A substation still exists close by, which itself was once fed by the generators at Bulls Bridge.

POWER STATIONS FOR TRANSPORT

Stockwell Generating Station

The City & South London Railway (C&SLR) was a pioneering underground line that set the standard for similar networks around the world. It was the first line on what would later become known as the London Underground to be a genuine subterranean railway. As discussed earlier, London's original

C&SLR electric train, with Stockwell Generating Station in the background. (© TfL, from the London Transport Museum collection)

Modest generators inside the plant at Stockwell. (© TfL, from the London Transport Museum collection)

'underground' lines were in fact built in trenches close to the surface, which could then be covered over once built. The C&SLR was the first to dig actual tunnels deep below ground, and one of the first lines to be dubbed the 'Tube'.

The company's other major contribution to London's transport system was that it was also the first to introduce electric locomotives. Electric traction provided a faster, cheaper and cleaner way of moving passengers than steam engines, and they would quickly become standard on the network.

The company was formed in 1883 under the name of the City of London & Southwark Subway. The necessary powers needed to start construction were granted in 1884, and work began in 1886. The initial project outlined plans for trains to be operated via a cable system that had been pioneered on the streets of San Francisco. But it soon became apparent that using cable haulage to move carriages over the full length of the line would prove to be slow and inefficient.

Plan B was to use electricity, which had recently been demonstrated elsewhere on minor railways as an effective way of powering trains. To lead the way forward as the first large-scale public railway anywhere in the world to use electric traction was an experimental risk, but the company was now committed, and initial tests showed encouraging results. The line opened in

1890 on a route which ran from King William Street station (now disused) near Monument, to Stockwell in South London.

The C&SLR was required to provide its own electricity in order to operate its trains, and so a small power-generating station was built next to the station at Stockwell. The same building also housed a train depot and maintenance works, uniquely designed part above ground, part below. This was so that trains requiring work could be moved from the line via a special tunnel, and then raised to the surface by a special lift, which was also powered by the electricity generators.

The power plant was a basic set up that included four steam-powered generators of limited capacity, each feeding power to the live rail that gave traction to the locomotives. The trains demanded so much power from the generators there was no excess capacity left over to provide electric lighting for the stations along the line, or even at the power station itself. Lighting was instead provided by gas.

The machinery at Stockwell was later updated, however, to increase power as the line became more popular. A number of pump engines were also later installed and used to provide hydraulic power for operating lifts at each station.

By 1913 the C&SLR had been absorbed into Charles Yerkes' Underground Electric Railways of London, and the generating station was decommissioned soon after in 1915. Power for the line was provided instead by the company's new power station at Lots Road, which fed into a small substation constructed close to the original generators. Stockwell Tube station was totally rebuilt in 1924, with the generating station, depot and maintenance works demolished.

The route was later merged with that of the Charing Cross, Euston & Hampstead Railway to form what is now much of the Northern line. Nothing remains of the power station building, but it was located close to the Tube station of today, at the junction of Clapham Road and Stockwell Road.

Wood Lane Generating Station

The success of the City & South London Railway's use of electric trains prompted other companies to follow suit, including the Central London Railway (CLR). It was granted permission in 1891 to build an underground railway that would run from the City, through the heart of central London and out towards the west.

After several years of delicate construction work that included tunnelling below some of London's most historical buildings, the line finally opened in 1900 with thirteen stations between Bank and Shepherd's Bush.

As with the C&SLR, the CLR had to supply its electricity in-house, and plans were drawn up before opening for a generating station to be built close

CLR engine in front of the Wood Lane Generating Station. (© TfL, from the London Transport Museum collection)

The station today, now White City Bus Garage.

to Queen's Road station, towards the western end of the route (Queen's Road station was later renamed Queensway). But with further extensions to the line already in mind, it was decided instead that the power plant should be built on a plot of land near to the terminus at Shepherd's Bush, close to Wood Lane.

Learning from the mistakes made at Stockwell, where the small size of the generating hall restricted the number of boilers it could hold, the CLR constructed two large brick buildings capable of generating enough power for the original line plus any future expansion. It was also able to provide lighting for every station along the route.

The line was extended in 1908 when a new station was built next to the generating station. Named Wood Lane, it was opened to serve the Franco-British Exhibition, and events running as part of the 1908 London Olympics that were taking place in White City.

Again mirroring the fate of the C&SLR, the Central London Railway was later absorbed by the Underground Electric Railways Company of London. They decided to close the generating station at Wood Lane in favour of power supply instead from Lots Road. The CLR route would later become a large section of what we know today as the Central line.

The generating machinery was removed after closure, and the building was initially earmarked for demolition. Instead it was sold to the Dimco tool manufacturing company, who renamed the former station as the Dimco Buildings.

Wood Lane Tube station closed in 1947 and lay abandoned for decades, as did the Dimco Buildings, which had also become derelict by this time. The decaying Tube station was finally demolished between 2003 and 2005 to make way for construction of Westfield shopping centre. The Dimco Buildings survived the demolition however; their beauty having been recognised with a Grade II listing in the early 1980s. They were fully restored in 2008 as part of the shopping centre complex, and today they are used as White City bus station. This brings them full circle and back to being part of London's transport network.

Lots Road

Lots Road Power Station is a huge disused building located at Chelsea Creek in south-west London. It often appears to be confused with Battersea, a mistake that is perhaps understandable considering the similarities. Both are mega-sized buildings that have sat abandoned for several years, with numerous failed attempts at redevelopment since. They are fairly close to each other, and even looked very similar until Lots Road had two of its four chimneys removed. But grouping the two power stations too close together doesn't do justice to the major role Lots Road has played in the development of modern London.

Lots Road Power Station. (© TfL, from the London Transport Museum collection)

Different to any other large power station in London that came before it, Lots Road was built solely for the purpose of powering underground railways, at a time when several existing companies were being amalgamated as part of the United Electric Railways Company of London (UERL).

It was originally the vision of a railway company founded in 1896 as the Brompton and Piccadilly Circus Railway (B&PCR), who planned to construct a new underground route from central London to South Kensington. A similar company known as the Great Northern and Strand Railway was created two years later in 1898, with a view to building a line that ran from Aldwych to Wood Green. The already well-established Metropolitan District Railway (MDR) was also looking to extend its network at the same time, part of which followed a similar route to the one being proposed by the B&PCR line.

The success of the City & South London Railway and Central London Railway made it clear that all three proposed lines would need to be electrified, but the costs this involved meant that each company struggled to find the necessary investment.

Charles Yerkes had already taken ownership of several underground railways in London by 1900 and was keen to expand his portfolio. He gave the MDR the finance required for their new electrified line, and later, realising the potential for combining them as one route, also purchased the other two companies. The MDR was kept separate, but the other two railways were combined to form the Great Northern, Piccadilly and Brompton Railway (GNP&BR), which today makes up a large portion of the Piccadilly line.

By 1902, Yerkes had created the UERL, and all three railways now came under its control. Instead of building separate generating stations to power individual railways, the scale of the planned site at Lots Road was expanded so that it would now be capable of powering every railway under Yerkes' charge, with enough capacity left over for any other lines that the company acquired.

Lots Road began generating in 1905, built on a scale never before seen, to a design by resident UERL engineer James Russell Chapman. It was the largest power station in the world on opening, and within thirty years it was providing power to the entire London Underground network, plus miles of tramlines.

This was a power station that literally moved London, helping millions of people travel around the city in every direction. Together with the Metropolitan Railway's power station at Neasden, Lots Road continued to

Lots Road Control Room. (© TfL, from the London Transport Museum collection)

provide for the Tube network with each new extension deeper into the suburbs, helping to create the commuter belt around London, and the development of the city as a sprawling metropolis.

The increase in demand meant that machinery had to be upgraded several times, including additional boilers added in the 1920s. The decline in coal-fired power stations also led to Lots Road being converted to oil power in 1969, although the increase in oil prices that led to the demise of Bankside Power Station prompted London Underground to convert Lots Road again the following decade, this time to gas. The removal of coal fires made four chimneys unnecessary, and so one of the original four was removed during the conversion to oil. This left the power station with something of an odd appearance, but the removal of a second chimney in 1979 restored the building to a more coherent look.

It was during the transition to oil that Lots Road also came to the assistance of the radio industry. New laws were introduced in 1972 that allowed commercial radio stations to legally operate in the UK for the first time. By the following year two new stations were ready to launch: the London Broadcasting Company (LBC), and Capital Radio. Transmission was to be via a new antenna tower in Hertfordshire, but complications meant it wasn't finished in time. The solution was to erect a makeshift antenna hung between two of Lots Road's chimneys, which allowed the two radio stations to launch successfully as planned. The power station was used several more times for occasional radio transmission by various companies until 2007.

By 1985, Lots Road was becoming increasingly expensive to operate, and it was decided that buying power direct from the National Grid would be cheaper. But due to concern that a major power outage would cause havoc on the London Underground and leave passengers stuck in tunnels, the generators at the power station were retained as back-up. The ageing machinery came to the end of its lifespan in 2002 and the entire site was finally closed, almost a century after opening (back-up power for the Tube network since then has been supplied by Greenwich Power Station – to be covered later).

A number of failed redevelopment plans have come and gone since the station closed, but at the time of writing Lots Road still stands derelict and empty. Its huge size and prime location makes it ripe for conversion to apartments, which would at least see the building preserved.

Today it can be seen in all its splendour along Lots Road itself, and by walking through the Chelsea Harbour complex. Except for partial demolition after closure, the building is in surprisingly good condition, with most of the brickwork and windows intact. Note that some of the windows were bricked up long before the eventual closure. It is believed that they were filled in during the Second World War as a way of minimising potential bomb damage. Lots Road did take a hit during enemy action, but it caused little destruction.

You can also get a great view of the station from Battersea Bridge, and from the opposite side of the river in the tranquil surroundings of St Mary's Church. It was from here that J.M.W. Turner painted the Thames some sixty years before the power station was built.

Neasden

If Lots Road Power Station was the Underground Electric Railway Company of London's way of stamping its authority on the growing underground railway industry, then Neasden Power Station was the Metropolitan Railway's response.

Despite years of ruthless competition with the Metropolitan District Railway – itself owned by the UERL by 1901 – the MR was still a powerful force on London's railways at the beginning of the twentieth century. The company was, however, losing passengers to the new 'Tube' railways, and so moves were made to convert the network from steam to electricity. This demanded the building of a new power station on a scale large enough to provide for their entire line.

A plot of land was chosen in the north London town of Neasden. It had risen in population since the MR opened a station there in 1880 as part of an extension from Baker Street to Harrow-on-the-Hill. It was characteristic of a company that would later create the concept of 'Metroland', where rail passengers were encouraged to buy cheap housing outside central London on land owned by the railway, thus creating suburban commuter towns.

The MR was fast building its own town-within-a-town at Neasden. By 1900 the self-styled Neasden Village included a train shed, maintenance depot and hundreds of houses for its workers, many of whom had been forced to move there after all maintenance work was transferred from a previous depot at Edgware Road.

The village provided its residents with schools, a church and various social activities, yet it proved to be unpopular with the wives and children of the workforce, and left them with feelings of isolation. Living in the shadows of the dirty, smoke-filled railway depot no doubt resulted in a low quality of life, perhaps not helped by the naming convention used for several streets in the village.

The list of road names included Quainton, Winslow, Verney and Aylesbury – each one a picturesque town or village in rural Buckinghamshire. The rationale behind the names was obvious: they were all places to which the MR had now reached, despite several being a considerable distance from London. But for the residents of Neasden Village, it possibly served as a constant reminder that they would rather be there instead, or indeed anywhere else other than

Neasden Power Station. (© TfL, from the London Transport Museum collection)

Neasden. Rather ironically, those same streets are still in existence today, some sixty years since the MR stopped serving these stations. To make matters worse for the families living in Neasden, the workers' homes had been built on poor land that later caused damage to several houses.

As a result, the addition to the village of a huge power station and the pollution that came with it was hardly a welcome development. The station was built regardless in 1904, and electrified services began operating on the line from Baker Street to Harrow a year later. Other parts of the company's railway were electrified later, in 1905, including parts of what are now the Circle and Hammersmith & City lines.

A fleet of more powerful trains were introduced in 1923 that led to Neasden being expanded to increase capacity. By now the electrification of the railway had reached as far as Rickmansworth, with every other part of the network being connected in the years that followed.

When the MR later merged with the UERL and various tram and bus companies to become London Transport, the power station at Neasden worked together with Lots Road to supply the entire underground network until it was decommissioned in 1968.

Former control room at Neasden. (© TfL, from the London Transport Museum collection)

It was demolished after closure and nothing now remains. The train sheds and depot do still exist and can easily be seen when travelling between Neasden and Wembley Park stations on the Metropolitan or Jubilee line.

The Buckinghamshire-themed street names can be seen leading off from Neasden Lane. The original houses have been updated and replaced, but the layout still provides some sense of what Neasden Village was once like. The entrance to the depot is located at the corner of Quainton Street and Chesham Street, but is not accessible to the public.

Chiswick Power House

At the same time as electricity was becoming established as the best way to power London's railways, it was also having an impact on its extensive tram and trolleybus network. Leading the way forward towards electrification was London United Tramways (LUT), a company that had been gaining huge ground since it was founded in 1894. It had been set up in order to amalgamate several small tram operators, with a view to updating existing infrastructure that could be run as one large-scale network. The company focused

The inside of Chiswick Power House in its prime. (© TfL, from the London Transport Museum collection)

on west London, and invested heavily in replacing horse-drawn trams with electric ones.

A dedicated power source was required to supply traction to each tram, conducted via overhead wires. This prompted the LUT to build a generating station at its head office and tram depot in Chiswick, west London. The station opened in 1901 and became known as the Power House.

It was designed in impressive style both inside and out. It was an approach that was typical in an era when even the most mundane municipal buildings were built to impress, usually as a way of making the owners look successful.

The Power House was the work of architect William Curtis Green, who had already designed two tram depots in Bristol for the LUT's parent company Imperial Tramways. Green would later go on to help design several buildings across London, including the Dorchester Hotel.

It was from the depot in Chiswick where the LUT's first electrified tram went into service in April 1901, with the first route running from Shepherd's Bush to Acton. Other routes were added soon after.

London United Tramways was absorbed by the Underground Electric Railways Company of London in 1913. Sharing the same fate as the generating stations at Stockwell and Wood Lane, the Chiswick Power House

London United Tramways stonework.

was decommissioned in 1917 when power supply for trams was handed to Lots Road. It was downgraded to a basic substation facility that was usually referred to by the far less glamorous name of Goldhawk Road (despite not actually being located on the road it was named after). The substation was also taken out of service in 1962, and part of the Power House was demolished a few years later in 1966.

A Grade II listing granted in 1975 has ensured no further demolition will take place, and today this attractive building can still be seen mostly intact. It is now a recording studio and office space, and can be found on Chiswick High Road, set back from the road in a courtyard. The exterior of the building includes a sculptured panel featuring an image of two women that are said to have represented locomotion and electricity. One of the original entrances also includes the initials LUET above the door, which presumably adds the word 'electric' into the company's name.

The control room equipment has been removed from inside, but some of the original iron staircases have been preserved. Further along Chiswick High Road is Stamford Brook Bus Depot. It is housed inside what was formerly the LUT's tram depot. It is not accessible to the public, but it is possible to see the Art Deco buildings from the street.

Greenwich

Greenwich Power Station is something of a forgotten gem. It often goes unnoticed by tourists as they cruise on tour boats headed towards historic Greenwich and its maritime glory. The building's imposing size makes it almost impossible to miss as it gets closer to each boat that arrives at Greenwich Pier, but any tour guide worth his weight in tips knows that here is where they need to tell passengers about the Cutty Sark, the Old Royal Naval College and the Royal Observatory.

The power station is located just beyond these historical sites, sitting quietly by the Thames, looking just as much abandoned as its more famous counterpart in Battersea. Except that Greenwich Power Station isn't actually abandoned at all, and is still in full working order.

In 1900 Greenwich officially became part of the newly formed Metropolitan Borough of Greenwich, which was governed by the County of London (both were later abolished and replaced by the Royal Borough of Greenwich). Similar to the London United Tramways (LUT) company based at the Chiswick Power House, the County of London owned a large tram network that was operated under the name of London County Council Tramways. Taking advantage of an Act of Parliament passed in 1870 that gave local councils the option to acquire the assets of existing tram companies after a set amount of time, the company had taken control of several different routes by 1900.

Next step was to electrify the new combined network and replace horse-drawn trams, but changes in local government policy led to the company having to adopt a different method of traction to the LUT. The London Government Act of 1899 resulted in the creation of new local boroughs that divided London into twenty-eight different districts. Each borough was granted power over the infrastructure in their area, and could therefore permit or deny a company the right to dig up roads for water, gas or electricity mains.

Crucially for tram companies, this also included provision for erecting overhead wires along streets. Whereas the LUT only needed to gain permission from a small number of west London boroughs; the council company had ambitious plans to expand, and would therefore need approval from many different areas.

It was decided that being denied permission for overhead cables in any borough was a risk they were not willing to take, and instead of overhead wires, traction was conducted via conduits in the road. It was also a cheaper option that involved simply adapting track that was already there.

The final element required was of course a power supply. Chiswick Power House was built on a scale similar to the generating stations at Stockwell and

Greenwich Power Station during its heyday. (Image used courtesy of the Greenwich Heritage Centre)

The disused but well-preserved coaling jetty at Greenwich.

Wood Lane. The County Council's power station would therefore need to be bigger and better, and construction started on the site in Greenwich in 1902.

On completion it became clear that this was a building of huge proportions, drawing comparisons to Lots Road which was built at the same time. The company's first ever electrified tram journey took place in 1903 on a route from central London to Tooting, and within twelve years horses had been phased out altogether.

The building itself is another design classic, continuing the trend for large power stations that began with Deptford West. It also shares similarities with the 'brick cathedral' style that would appear later in the power stations of Giles Gilbert Scott.

But if Battersea and Bankside compare favourably with the heavenly overtones of St Paul's, then Greenwich definitely veers more towards the hellish end of the spectrum, with a dark and gothic-looking exterior. It was a design that no doubt blended well with the other dirty power stations and gasworks close by.

Similarly to what would later happen at Battersea, construction of Greenwich was delayed by concerns about height. The complaints were not from the Archbishop of Canterbury this time however, and instead came from a group of men with a different celestial concern: the astronomers of the Royal Observatory. In order not to interfere with their stargazing, it was agreed that two of the power station's four chimneys would be shorter than the other two. This allowed for a clearer view of the night sky.

London County Council Tramways became part of the newly formed London Passenger Transport Board in 1933. With the Underground Electric Railway Company of London also now part of the same group, the power stations at Lots Road and Greenwich were combined to create a joint output large enough to run the entire underground railway and all of London's trams.

Greenwich was converted to gas-fired operation in the 1960s, which was seen as a more efficient method of generation. The original complex included a substantial coal jetty on the Thames, which was rendered obsolete once the switch to gas was complete.

London Underground's decision to transfer responsibility for power supply to the National Grid led to the closure of the ageing Lots Road. But instead of closure, Greenwich was upgraded with new machinery, and is now used as a back-up facility for both the National Grid and London Underground. The equipment is tested on a regular basis, and current owners UK Power Network Services claim that the station can be ready for action in just fifteen minutes.

The power station can easily be seen from many streets in the area, including Old Woolwich Road and Hoskins Street. You can also get even closer by walking along the Thames Path, which allows good views of the large coaling jetty, which is completely intact.

The oil generators installed in the 1960s were considerably smaller than the four giant steam turbines installed originally. This means that today the inside of the power station is a cavernous empty space where very little machinery remains.

Stonebridge

The introduction of electric locomotives on London's underground railway also started a gradual move away from steam engines on the city's main-line rail network, and one of the first major companies to make the transition was the London & North Western Railway (LNWR). The company was one of the largest and most successful at the start of the twentieth century, having acquired various previous lines since it was founded in 1846. It was a railway that covered huge parts of the United Kingdom, with important routes that connected London to Manchester, Birmingham, Liverpool and other large cities.

The company began an ambitious project in 1909 to electrify the bulk of their suburban lines within London, specifically the routes between Watford Junction and Euston, and Richmond to Broad Street (Broad Street station closed in 1986 when its last remaining services were moved to Liverpool Street nearby). The aim was to improve journey times with faster trains, and to ease congestion on the railway by separating services into slow, local routes and fast, inter-city ones.

The plan also involved building a connection to the Baker Street & Waterloo Railway, which had opened in 1906 on a route that ran underground from Lambeth North to Baker Street. It was renamed the Bakerloo line months after opening, and was soon expanded towards Paddington. From there a further extension was planned as far as Queen's Park, and it was here where it would connect to the LNWR. Progress was hampered by the start of the First World War, but electrified services were finally introduced in 1914. The connection to the Bakerloo line at Queen's Park opened the following year, and the route between Broad Street and Richmond was also ready to begin operation by 1916. Combined, the improved railway was the biggest network anywhere in Britain to become fully electrified, setting the standard for every other major rail company operating in London.

The large amounts of electricity required to run the routes prompted the LNWR to build a dedicated power station in 1913, next to their railway station at Stonebridge Park. It was an extensive operation that was almost on a par with the facilities at Lots Road and Neasden, with coal delivered via the main line.

The Railways Act of 1921 led to the company being absorbed into the London Midland and Scottish Railway, which became part of British Railways on nationalisation in the late 1940s. Stonebridge Power Station con-

Stonebridge Power Station. (© TfL, from the London Transport Museum collection)

tinued to operate nonetheless, surviving a bomb attack during the Second World War. Its demise came in 1967 when responsibility for electric power on Britain's railways was handed to the National Grid.

Today, trains still run along the original electrified route, including London Overground services between Euston and Watford. The line is shared with Bakerloo line trains between Queen's Park and Harrow & Wealdstone. Both operators still stop at Stonebridge Park, but the power station building has long since been demolished. It was located on what is now the Bakerloo line's Stonebridge Park Depot. This can easily be glimpsed from passing trains.

Great Eastern Electric Light Generating Station

Another powerful rail company operating at the turn of the century was the Great Eastern Railway (GER). Formed in 1862, the company focused on a number of key routes into London from the eastern counties. Its original terminus was at the long-since demolished Shoreditch station, located on Shoreditch High Street and later renamed Bishopsgate.

In 1874 the company built a lavish new terminus at Liverpool Street, on a scale grand enough to match the great termini built by their competitors at St Pancras, Kings Cross and Victoria. The company decided to enhance the station with electric lighting in 1893, replacing the gas lamps that had lit the

Great Eastern Electric Light Generating Station.

platforms and concourse until then. The new improvement required the GER to produce its own electricity supply, and so a generating station was constructed in nearby Shoreditch, close to the former terminus at Bishopsgate.

It is questionable why the GER invested in a generating station simply to power lighting at a time when other London railways were making plans to electrify their actual lines. The company instead focused on steam, placing them far behind the other major railways. The generating station continued in operation until 1932, by which time the Railway Act of 1921 had absorbed the GER into the London and North Eastern Railway. All machinery and equipment was removed in 1934, and part of the building was demolished to remove its chimney.

The rest of the stylish red-brick building is still standing proud today, and can be seen at 233 Shoreditch High Street. It was restored in 2009, but then faced demolition soon after when plans were announced for a new commercial development on the site. A campaign to save the building led to the developer amending its plan so that the generating station could be maintained; however, the entire project appears to have stalled in recent years as a result of the recession. The building is now home to a bar and restaurant appropriately named The Light, giving visitors the opportunity to venture inside.

Baker Street Substation and Others

The transition from electricity being supplied to the London Underground via the National Grid instead of dedicated power stations has left a number of substation buildings becoming largely abandoned. Almost all of them are still used in some capacity, but on a far smaller scale than previously. The legacy is a collection of buildings whose large size no longer matches the importance of their function.

Baker Street station is one of the oldest on the entire Underground network, and was part of the original Metropolitan Railway line from Farringdon to Paddington, opened in 1863. It was built to a lavish design from day one, and even more so when the entire station was rebuilt in the 1920s. It was also during this period that a large substation building was constructed further along Baker Street at number 228, which was fed from Neasden Power Station.

Today the building is unnoticed thanks to a row of tired-looking cafes and shops that now inhabit the ground floor. Look above, however, and you'll see a wonderful building with impressive stonework and windows reminiscent of a church. It was designed by the Metropolitan Railway's Chief Architect Charles W. Clark, built on a plot of land where once stood a house owned by famous eighteenth-century actress Sarah Siddons. It was where she died

Substation building at Baker Street. (© TfL, from the London Transport Museum collection)

Inside the substation at Wood Green. (© TfL, from the London Transport Museum collection)

in 1831, and her spirit is said to haunt the substation building that replaced her home. Siddons' memory lives on today in the form of a preserved electric locomotive that the MR named after her. It was in service until 1961 and has been restored to working order by the London Transport Museum.

Another large substation building can be found close to Wood Green Tube station in north London. The station opened in 1932 as part of an extension to the Piccadilly line, and was designed by renowned Tube station architect Charles Holden. The substation building can be found further along High Road from the Tube station, opposite Wood Green Bus Garage. It is easily recognisable as a large brick structure with narrow windows, similar to the look of many of Holden's Tube stations. It is unclear how much equipment is left inside today, but the substation formally housed a sophisticated control room to assist with traction supplied from Lots Road. The Tube station is Grade II listed, but it is unknown if this includes the substation.

Another distinctive Charles Holden Tube station can be found in Northfields in west London. It was built in 1908 on a westerly extension to the Piccadilly line that would eventually reach Heathrow. Similar to Wood Green, a substation was built close to the station in order to receive power from Lots Road. It is another imposing, tall, brick building almost identical to the one at Wood Green, and it can be seen by walking along Redwood Grove, close to the entrance of Northfields Depot.

Northfields Substation. (© TfL, from the London Transport Museum collection)

Battersea Power Station.

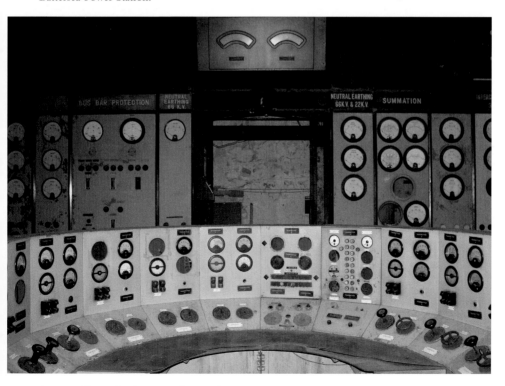

Battersea Power Station control room A.

The abandoned Lots Road Power Station as it looks today.

Chiswick Power House.

Gasholders of Vauxhall Gasworks, next to The Oval.

The former gasholders of Nine Elms Gasworks, close to Battersea Power Station.

Greenwich Power Station.

Lots Road Power Station.

Office block at the abandoned Bow Common Gasworks.

Bromley-by-Bow Gasworks.

Imperial Gasworks.

Wapping Hydraulic Power
Station.

Limehouse accumulator tower.

Abbey Mills Pumping Station as it looks today.

A concrete base is all that remains of one of the two demolished chimneys at Abbey Mills Pumping Station.

Imposing chimney stack at Kew Bridge Pumping Station.

4

BRINGING GAS TO THE PEOPLE

GASWORKS FOR INDUSTRY AND PUBLIC SUPPLY

The huge demand for gas in the nineteenth century led to scores of gasworks being built across London. Some were small, modest operations. Others were larger than whole towns. They employed hundreds of thousands of Londoners, but their demise came as a consequence of industrial changes and the discovery of cheaper ways to obtain gas.

The result was mass closure of all of London's gasworks, with most sites demolished and cleared. The only structures usually left standing were large gasholder frames, which even today are a common feature in the skyline of any major UK city. The immense size of London and its surrounding suburbs means that the metropolis is peppered with these weird, wonderful and often highly decorative structures. They can be seen across all boroughs, from the poorest districts to some of the wealthiest, and virtually every famous view of the city has at least one gasholder frame lurking somewhere.

The gasholders may be all that remains of most of the works covered here, but they were once just one of several buildings included at each site. From offices and showrooms to retort houses and coal sheds, each played a vital role in the manufacture of gas for use by millions across London.

The story of how each was built and run is filled with companies with grand names, the ruthless competition between them, and how each gasworks affected the lives of those who worked and lived in and around them.

But first, it is necessary to take a brief look at exactly how a gasworks operated in nineteenth-century London and beyond. Similar to the power stations described earlier, the gas production process began with coal.

Furnaces were used to heat the coal inside what was known as a retort. Early gasworks included horizontal retorts made from iron. These were gradually replaced with vertical ones made instead from clay, which proved to be more robust. The vertical retort was developed by Frederick Albert Winsor, son of the great founder of the Gas Light & Coke Company. Retorts were typically housed inside a large brick building. Many individual retorts were needed, and therefore the 'retort house' was often one of the biggest parts of a gasworks.

The heating process created gas vapours, which would then be extracted from the retorts and cooled. The gas then had to be purified in order to remove unwanted substances and chemicals. If left untreated the gas would be less effective for the consumer, and also potentially hazardous to health. The purification process included the use of lime to filter the unwanted waste, although this method was later dropped in favour of using water jets.

After purification the gas was ready to be distributed to the consumer, and was stored inside gasholders until needed. They were used as a way of ensuring a gasworks always had enough for a constant supply, achieved by manufacturing more than what was necessary at any one time.

Early gasworks were pioneered by gas industry innovator Samuel Clegg, and gradually began to be constructed on a larger scale as the industry and demand grew in unison. Most early gasholders were rigid, but later variations included telescopic levels capable of rising up and down, dependent on how much gas needed to be stored.

Each holder consisted of a huge tank, where the gas itself was held, and a column-shaped metal guide frame. It is these metal structures that can still be seen across the city today.

When the gas inside a holder was required, it was delivered via mains pipes built underneath roads. One large pipe would typically cover an area, with smaller pipes branching off to serve individual streets. It was from these mains that gas could then be connected via pipes to actual homes and businesses.

In this chapter, the gasworks described were all built in the 1800s for the manufacture of what became known as town gas (also sometimes referred to as house gas). This differs from natural gas, which was discovered in the North Sea and led to the decline of gas production.

Most sites listed here have gasholders still in situ. A distinction is made where necessary between a gasholder and a gasholder frame. 'Gasholder' is the more accurate term, although they are also often referred to as 'gasometers'.

Many gas companies of the nineteenth century had long and similar-sounding names. To avoid confusion, each one is listed under its full title the first time it is mentioned in a section or chapter, and then shortened thereafter.

Shoreditch and Haggerston

There are at least four former gasworks in Shoreditch and Haggerston, east London. Although completely separate sites owned by different companies, they were located very close to each other. Two in fact existed almost side by side, and because of their close proximity, they are discussed here in one entry.

Original wall remains at Shoreditch Gasworks.

Shoreditch was already a key stronghold for the early gas industry by 1820. It was here, in 1813, where the Gas Light & Coke Company built one of their earliest gasworks on Curtain Road. It was located close to Norton Folgate, and built in order to serve this small but historic area on the boundary with the City of London. As discussed earlier, however, the site chosen on which to build the works had been purchased in haste, and proved to be unsuitable. The rush had been to ensure that the lucrative Norton Folgate contract did not get handed to a rival company, but delays in construction caused by poor workmanship meant that the company ended up losing the contract anyway. It was a less than successful start for Curtain Road, and it would soon get worse.

Complaints from local residents about the smell of tar put pressure on the company to try and improve its working practices. They were also later reprimanded after it was discovered that tar residue was being taken away from the works via pipes and dumped directly into the Thames, at a time when the river was London's main source of drinking water.

The poor planning of the site also meant that it was increasingly expensive to run. The biggest issue was that Curtain Road was nowhere near a source of water. Coal could therefore not be delivered by barge or collier, and instead had to be transported to the works by road – a far less cost-effective method of delivery than water.

The hazardous process of producing gas in the early nineteenth century also sometimes led to industrial accidents, which generated significant fines and compensation charges for the company each time a surrounding building was damaged.

It was for these reasons and more why as early as 1826 the works were being recommended for closure. However, none of the other works yet built by the Gas Light had enough capacity to take on the extra demand that closing Curtain Road would create, not even their recently built other Shoreditch works close by on Brick Lane.

With full closure therefore not an option, Curtain Road was instead scaled down so that it would only be used during the winter months when demand for gas was higher. It finally closed for good in 1870 and the entire site was later demolished. Nothing remains today.

Further north from Curtain Road, another Shoreditch Gasworks was built in the 1820s by the Imperial Gas Light & Coke Company, who had been granted the powers needed to start providing gas in 1821. The company chose a site in Hackney Fields, a name which originated a century earlier when Hackney was largely farm and marshland. The building of the Regent's Canal had brought much industry to the area, and it was the closeness to water that had attracted the Imperial Company. They were no doubt keen to avoid the same mistakes as at the badly placed Curtain Road works of their great rival.

Shoreditch Gasworks later became part of the Gas Light's network of sites when it amalgamated with the Imperial Company in 1876, and in 1901 the new owners set about updating the works in order to increase output.

By 1934 the Gas Light had several large gasworks in operation, leading to the decline of several of their various smaller works, including Shoreditch. Just as with their own works at Curtain Road decades earlier, it was downgraded to the status of a stand-by facility. For the rest of its lifespan it was only called into action if demand was exceptionally high during the cold winter months.

Parts of the works were then destroyed in 1944 when the site took a direct hit during the Second World War bombing campaign. The cost of repairing the damage was deemed unfeasible considering its low output, and the works were finally closed in 1949. The derelict remains of the site were demolished in the mid-1950s, and in 1958 it was redeveloped as public parkland that is known today as Haggerston Park. It can be accessed via Audrey Street or Goldsmith's Row, and includes small traces of the original gasworks boundary wall. The corner of the park nearest to Whiston Road and Dove Row is where the works' gasholders were formerly located.

Further along the canal from the works at Shoreditch was Haggerston Gasworks. It was built by the Independent Gas Light and Coke Company, who began operating in 1829 and within a few years had become a serious threat to the dominance of the Gas Light.

Gasholder frames at Haggerston Gasworks.

As at Shoreditch, the canal-side location allowed for coal to be delivered to the works with ease, and in fact Haggerston works was built even closer to the water than its nearby rival, literally on the towpath.

The Independent Company later became another of several companies absorbed by the Gas Light in 1876. Production continued here until the turn of the century when it was decided that the Haggerston works were now surplus to requirements. Final closure came in 1908.

A portion of the site was converted into a showroom after closure, where gas appliances were demonstrated to the public. It was an idea that had been introduced at several other sites owned by the Gas Light, where female workers were used to sell products in a somewhat desperate attempt to prevent consumers from using electrical appliances instead.

Nothing remains today of the showroom or other office buildings, but two elegant and decorative Victorian gasholder frames from the original works still stand alongside the canal. These can be viewed by walking to the entrance gate of the gasholder site (known today as Bethnal Green Holder Station) on Marian Place, or from the end of Darwen Place. They are best seen, however, from the opposite side of the canal on the towpath along Andrews Road. It is from here that the ornate metalwork can be appreciated in full, a testament to the skill of the engineers of Victorian London in being able to make the most unglamorous places and structures a thing of beauty.

Pancras

The Kings Cross area of London is synonymous with London's railways, including two major railway termini and one of the busiest Tube stations on the entire London Underground network. The first railway station, Kings Cross, was opened here in 1852 by the Great Northern Railway. This was followed in 1868 by St Pancras, built by the Midland Railway. The two stations changed the face of the area forever, taking over a huge plot of land that also included a vast complex of sidings and goods sheds that were located behind the main station buildings.

Most of the goods sheds had been closed by the mid-1970s, adding further dereliction to a site that was by now a wasteland. It had every single obligatory characteristic of a typical abandoned industrial landscape: decaying buildings, disused train tracks and sheds, empty factory buildings with smashed windows and graffiti, and, towering above everything else, a collection of old Victorian gasholders.

Pre-dating the coming of the railways by almost thirty years, the metal gasholders that contributed to the bleak look of Kings Cross in the 1970s were all that remained of what was at one time Britain's largest gasworks. They were opened in 1823 by the Imperial Gas Light & Coke Company in order to provide gas lighting to various local businesses and homes. Named simply Pancras Gasworks – the 'St' was dropped for reasons unknown – they were the result of permissions granted to the company in 1822 that allowed them to build a large works on the site.

They were to be the Imperial's way of confirming its arrival as a major competitor to the Gas Light & Coke Company, and were therefore built on an unprecedented scale. The new works were designed by the company's in-house architect Francis Edwards, and construction began with the first foundation stone being laid by Sir William Congreve.

Mostly remembered for being the inventor of a series of early military rockets, Congreve had also conducted several gas experiments that proved to be influential in the development of the gas supply industry. He had also been involved at some time or another with first the Gas Light and now the Imperial, and was therefore more than qualified to be the guest of honour at the Pancras ground breaking ceremony.

Actual production began at the works in 1824, and within a few months it was supplying to hundreds of buildings across what was then the Metropolitan

Disused gasholder frame at Pancras Gasworks. (Image used courtesy of Paul Talling)

Borough of St Pancras (absorbed in 1965 into what today is the London Borough of Camden). An explosion is said to have caused significant damage to several buildings at the works in 1826, but it was reconstructed quickly, and returned to full capacity soon after.

Much of the success of the works hinged on its location next to the Regent's Canal. The canal's first section had opened in 1816 between Paddington and Camden, and by 1820 it had been extended from Camden all the way to the Thames at Limehouse. It was a stretch that ran through St Pancras and Kings Cross, and a special basin was built as part of the gasworks in order to allow coal to be delivered by barge.

The site included the most up-to-date gas production machinery on opening, but was further modernised to increase capacity in the early 1860s. It was at this time that the first set of steel gasholder frames were constructed. Several more were added over the next three decades, and they would come to dominate the local skyline for over a century.

With ornate decorative touches and a layout that seemed as though it was designed by an artist who knew the shapes they created would one day be admired from above, it was clear that the Imperial wanted its collection of gasholders to impress. Of particular beauty were three of the holders nick-named the 'Siamese Triplets'. They were built very close to one another.

So close in fact that they shared certain interlocking sections, giving the impression of three circles connected as one – much to the excitement of architectural critics years later.

In reality the Siamese Triplets were not the result of artistic creativity. They were designed this way merely to maximise the limited space available. Although the Regent's Canal had brought many benefits to the gasworks, it also left the Imperial with little room for expansion. Space became even more of an issue by the latter part of the nineteenth century, as by now the railway stations close by had become the new major landowners in the area, making the gasworks site even more cramped.

The construction of the many gasholders also caused uproar from local residents, most of whom complained that they would devalue their houses. Additional holders were added regardless, and the earlier ones were updated to make them telescopic. They were now able to rise and fall as demand required, and many other gasworks also began to install telescopic holders in the following decades. But the constantly changing appearance of the perceived eyesores as they moved up and down further exacerbated local homeowners.

Amalgamation led to the Imperial being absorbed into the Gas Light in 1876. It was to be the beginning of Pancras Gasworks' demise, as the new owners began to close several of the works it had acquired by taking over the assets of many smaller companies. The Gas Light instead focused their attention on their major works in East Ham (see later section), but Pancras was spared full closure in the immediate years following the takeover.

It was downsized instead to a works used only when demand was exceeded elsewhere. Full closure came in 1904, and the site was cleared three years later in 1907. Several of the gasholder frames were retained, however, among them the Siamese Triplets, and holder number 8, which was one of the most decorative of the collection.

What was left of the gasworks then sat derelict for several decades before planning permission was granted in 2006 for the entire area of land behind Kings Cross and St Pancras to be redeveloped in a project known as Kings Cross Central. Sensitive to the plight of several historical buildings on the site, the developers have refurbished many of the original structures from the former train yards. Some have also now been granted Grade I or II listed status, including the Siamese Triplets and gasholder 8.

At the time of writing the holders had been removed from the site in order to be restored. They will be rebuilt almost at their original location, and repurposed for residential and leisure use – see chapter 7 for more information.

Until the gasholders are returned home, you can see where the gasworks were located on the open space to the right of Goodsway, adjacent to

St Pancras International. There were entrances located on what has since been reconfigured as York Way, and on Battle Bridge Road, a street which still appears on maps but seems to have disappeared as part of the redevelopment work. It may also be possible to spot traces of the former coaling wharf as you walk along the towpath of the Regent's Canal.

You can also get a wonderful sense of what the former gasworks were like through the work of local artist and writer Angela Inglis. She has spent several years documenting the area, including the gasholders.

Stepney

In his novel *The English Monster*, author Lloyd Shepherd tells a fictionalised version of the real-life Ratcliff Highway murders. One of the grizzliest crimes ever committed in London, seven people were brutally killed in their own home, including a young baby. The murders took place in 1811 in Wapping, along a stretch of the infamous Ratcliff Highway, which Shepherd accurately paints as perhaps the most depraved and crime-filled road in London. North of Wapping, the road passed through Limehouse and close to Stepney, both of which were becoming heavily industrialised at the time of the murders.

Much of the route still exists today as simply the Highway, and one of the streets leading off it near Stepney is named Schoolhouse Lane. It was here that a small gasworks was built by the British Gas Light Company in 1825.

Gasworks wall remains at Stepney.

Founded the year before, the company caused some concern for several existing gas firms when it became clear that this new competitor was planning to supply to the lucrative City of London.

Another new company also set up shop in 1824 with a gasworks close to the British Company site, located just off the Ratcliff Highway on a piece of land known as Sun Tavern Fields (near to what is now King David Lane). It was built by the Ratcliff Gas Light & Coke Company, who later absorbed a rival gas company in Wapping and made it their headquarters.

The British Company and Ratcliff Company would both prove to be unsuccessful. The former had been forced to abandon their ambitious plans to advance towards the City, deciding to focus instead on east London suburbs such as Limehouse and Bethnal Green.

The Ratcliff Company, meanwhile, found it difficult to attract customers in the face of stiff competition from the larger, most established companies. Both companies were later absorbed by the Commercial Gas Light & Coke Company, who by the late 1830s had become a formidable player in London's gas supply hierarchy.

The two works already built in the area made it clear that this particular part of east London was the perfect place in which to build a large-scale operation, and in 1839 construction began on what would become the Commercial Company's Stepney Gasworks.

The plan for the new site was masterminded by Chief Engineer Isaac Mercer, who successfully convinced the board of directors that the works would need to be built as large as possible if they were to make a serious play for dominance over east London and beyond. As discussed earlier, the arrival of the Commercial Company had been met with anger from many existing firms. Those such as the British Company had not been averse to using sabotage tactics to try and ruin the new company's reputation.

The Commercial Company was able to brush aside such distractions however, and quickly began to thrive. By 1846, just two years after becoming fully operational, Mercer was asked to draw up plans for how the works at Stepney could be expanded in order to allow for a higher production rate. The engineer delivered a series of grand enhancement ideas, and the extension works began soon after. Several other expansions would follow, each one the result of greater demand and continued success. It would eventually grow to become one of the biggest amongst many gasworks in east London, with multiple gasholders that were visible for miles around.

Stepney Gasworks also included its own internal railway, built in 1912 for the movement of coal. Up until then the works had been supplied coal via barges at its own custom-built jetty alongside the Regent's Canal. Using large collier ships was far more economical, however, and so a wharf was used by

Stepney gasholder monument.

the Thames in Wapping, from where it was loaded onto trucks. It was then driven to Stepney and distributed around the works as needed by use of the internal railway.

Similar to later mega-sized works at Beckton and East Greenwich, part of the site at Stepney was dedicated to the processing and selling of chemicals resulting from the gas production process. Other parts of the works included the Commercial Company's head offices, and an extensive showroom where gas-related products were demonstrated to consumers.

The end for Stepney came in 1945 when gas production was ceased. It was officially closed a year later, with nationalisation of the entire industry following shortly after. The largely derelict remains were demolished in 1952, although several of the gasholder frames were retained for storage purposes for another forty years.

These were also cleared however, when the site was redeveloped as a new housing estate opened in 2005. It was a much-needed project that had been championed by several local groups who demanded new homes be built on the site, while still maintaining its importance as a historical industrial location. When permission to build the new estate was granted, the construction company had to go through a costly but necessary decontamination process to ensure the land was safe enough for residential use.

There are a few traces of the gasworks still left today, including a short length of original wall that can be found close to the canal, on Ben Jonson Road. In accordance with the local residents' group's wishes, the designers of the new estate have also paid tribute to the lost gasworks with a quirky monument that can be seen along Harford Street, at the junction with Dongola Road. At first glance it appears to be four black columns, each with a stone base. Closer inspection reveals that they are in fact restored sections of one of the original gasholder frames. They have even been arranged in a partial circle, creating a simple but effective visual link to what once stood here.

The legacy of the gasworks has also been recognised in the names of several streets within the new housing estate, including one named Mercer Court – a reference no doubt to the Commercial Company's former engineer.

The remains of the canal-side coaling jetty were still visible in the 1970s, but it has been removed as part of the redevelopment work. When standing on the towpath from the other side of the canal to the former gasworks site, it is possible to see some very faint markings where the jetty once sat.

The Ragged School Museum is close by, originally built in 1877 by Thomas John Barnardo as a school for poor children in the area. Taking into consideration its close proximity to the gasworks, it is likely that many of the children who attended the school had fathers who worked there.

Poplar

Poplar is now lost somewhere in the shadows of Canary Wharf, passed through unnoticed by thousands of people every day on the Docklands Light Railway (DLR). But it was once at the heart of the finest docks in the world, in particular the West and East India.

It has also been home to two different gasworks. The first was built in 1821, opened by a company known as the Poplar Gas Light and Coke Company. It was set up by a private group of investors for the sole purpose of building the site so that it could then be sold, ready-made, to an established gas company. They found a buyer in the shape of the Poplar Vestry, who had been swayed in their decision by pressure from local residents desperate for street lighting to come to the area.

The new works proved to be unsuccessful, and in 1840 a high-ranking official from the Gas Light & Coke Company was sent in to try and turn things around. Not even the assistance of the biggest player in the industry could make it work however, and the only solution left was to sell the works onto someone else.

It was purchased this time by the Commercial Gas Light & Coke Company in 1850. They were eager to expand their business beyond the major works

One of the remaining
gasholders of Poplar
Gasworks.

already established at Stepney, but just two years later in 1852 they decided to
close down the new acquisition in Poplar. It was later replaced instead by a
new works built a short distance away on Leven Road.

This was to become the second Poplar Gasworks, and opened in 1878.
Together with Stepney, the new works gave the Commercial Company even
greater sway in east London, one of the few areas in the whole of the city that
the Gas Light was never able to fully dominate.

The gasworks at Poplar continued to manufacture gas until the late 1940s
when production ceased. The nationalisation of the industry, and the long list
of private companies that disappeared as a result, led to Poplar being used only
for the storage of natural gas.

Today, one of the surviving three gasholders still appears to be used as storage. The other two are well-preserved metal holder frames that can easily be
spotted from Leven Road. Maps from the time when Poplar Gasworks was
in full operation show a significant number of buildings on the site between
Leven Road and the Lee Navigation canal, including a private wharf. All of
this has since been demolished. The gasholders can also be seen from the
other side of the canal at Cody Dock, not far from the former Bromley-by-
Bow Gasworks site.

Nothing remains of the original works, but they were located close to what
is now a small road named Ming Street, overwhelmed somewhat by the roar
of the West India Dock Road elevated above it.

Fulham

The most remarkable thing about the old Imperial Gasworks at Fulham is that they are still having an impact on the local area today. Considering its location so near to Chelsea, you perhaps would have expected developers to brush over the fact that this entire area of Sands End was once dominated by the works. Instead, the legacy has actually been embraced by developers, with several luxury apartment complexes built in the shadows of the surviving gasholder frames.

The former works are even referred to in the name of the large area of redevelopment along the Thames nearby, and the railway station that was built to serve it in 2009. Whether or not the people who live there and use the station understand the reference is something that remains to be seen, but there's no doubt that naming it Imperial Wharf is in recognition of the old works. Much of the works itself has even been converted for use as office space under the simple name of the Old Gasworks.

As the name suggests, the works were built by the Imperial Gas Light & Coke Company in 1824. The piece of land chosen was part of the former Sandford Manor, which was once a sprawling estate that included a manor house said to have been the home of Nell Gwyn.

On opening, the site consisted of a basic collection of gasholders that stored gas supplied from the company's existing works at St Pancras. But by 1829 it

One of the surviving buildings at the Imperial Gasworks, Fulham.

was decided that Fulham should also manufacture, and as a result the entire site was expanded with the necessary plant and buildings needed to produce gas.

This did, of course, necessitate the need for coal – a factor not considered when the location was originally chosen. A wharf was constructed close by at Chelsea Creek, adjacent to an area of land that would later be home to Lots Road Power Station. The building of the wharf proved troublesome, however, due to disputes with the canal company who owned the creek.

Also causing frustration was a new railway that had recently been constructed on land that ran between the gasworks and the wharf (the line is used today by London Overground trains that serve the station at Imperial Wharf). Production was therefore delayed by several months, but the works were finally up and running by 1830, and were soon supplying gas to consumers across much of affluent west London.

Although many of the wealthy residents and businessmen living in the area were no doubt pleased to have a gas supply, one particular man living close by was far from pleased that gas production had come to Fulham. His name was Charles Random de Berenger, an elusive character with a reputation as one of the most infamous scoundrels of the early nineteenth century. His colourful past included serving time in prison, being part of a fraudulent stock market scam during the Napoleonic War and marrying a rich baroness.

The marriage awarded him the title Baron de Berenger, which he used as leverage to become a member of London's high society. He had acquired enough wealth by 1830 to purchase a mansion known as Cremorne House on Lots Road, but was horrified by the opening of the gasworks and the foul smell it created. He made his voice heard to the Imperial shareholders but to no avail, and it is possible that his disgust at the gasworks was one of the reasons why de Berenger abandoned the house and the pleasure gardens he added to it.

Similar to Fulham Power Station that would later be built close by; the gasworks made a huge impact on the community of Sands End and provided jobs for thousands of workers. They were well looked after by the Imperial, with cottages built near the works for staff to live in.

The workforce was also taken care of in 1886 when many suffered from an outbreak of smallpox. By this point the site was now under the ownership of the Gas Light & Coke Company, and senior staff called upon the might of their new owners to overturn a decision by the local council to open a temporary hospital at the gasworks in order to treat those stricken by the illness. Fearful that they too may contract the disease, healthy workers and local residents objected to the hospital, and the Gas Light was successful in having it removed.

Imperial amalgamated with the Gas Light in 1876, and was one of the more significant companies they acquired. The Imperial was the biggest and most well established of all the many companies that were absorbed by the Gas Light, and with large works at Beckton to the east, Pancras to the north, Fulham to the west and soon Nine Elms to the south, the company now had a stronger hold than ever on London's gas supply industry. They expanded the works at Fulham with modernised machinery and additional gasholders, and also rebuilt the wharf to allow for even larger collier ships to be used for coal delivery.

The decline in production at Fulham came in the late 1960s as manufactured gas fell out of favour. The size of the workforce had been cut dramatically in the 1920s as the works returned to being used primarily for storage rather than production. Three of the gasholders are still used for this purpose today.

The works at Fulham are perhaps the most well preserved of all London's lost gasworks, and not simply because part of the site is still used for storage. Its redevelopment as the Old Gasworks means there is still much to see today, and the entrance can be found off a small roundabout between Harwood Terrace, Michael Road and Waterford Road. Entering the complex allows you to walk through what was once the main entrance to the works, and from here it is easy to spot the ornate gasholder frames still intact.

Heading towards the main gasholder also allows you to see various other buildings now used as office space or as studios by artists. Two older looking buildings mark what was once the Imperial's laboratory and offices, and both earned Grade II listed status in 2007. The laboratory was designed by architect Walter Tapper, renowned for designing churches across the UK, and also for working on some of the buildings in the grounds of Westminster Abbey.

Close by can be found two war memorials, the larger one of which is also Grade II listed. They were erected by the Gas Light to commemorate workers at the site who lost their lives in the First and Second World Wars.

The largest gasholder frame can be seen in all its beauty at the end of the road running through the works. It was constructed in 1830 and is regarded as the oldest gasholder still in existence anywhere in the world. It was awarded Grade II listed status in 1970.

There is also much to be seen outside the works site, including a quiet residential area nearby known as Imperial Square. It is a collection of cottages that were built by the company between 1874 and 1894 for its workers and those recently retired. Today it has the feeling of being like a small village, hidden away from the rest of Fulham. It was also here where immigrant workers from Germany are said to have lived when they were called in to replace men from the gasworks during a strike in the 1880s. It led to the square earning the nickname 'German Square', and it is claimed that Emden Street, from which the square is accessed, was named in honour of the town in north-west Germany from where the workers came.

Emden Street leads to Imperial Road, where the great gasholder frame can be seen in more detail. It is also possible from here to view the gasworks in context with the two chimneys of Lots Road Power Station in the background. Also along Imperial Road can be found a row of single-storey industrial units, known as Imperial Studios. They were originally part of the gasworks site, but were sold off as office space during the 1970s, when the works was downgraded to be used for gas storage only. Cremorne House is long gone, but its former location is now Cremorne Gardens, and can be

Office block at the old gasworks in Fulham.

visited on Lots Road, near to the former power station (covered in an earlier section).

Note that the naming of the gasworks, and the fact that it is a name still being referenced today, is itself a curiosity. To name them the Imperial Gasworks is fairly generic, especially considering they were not even the company's first site. Calling them simply Fulham Gasworks would have been the more logical choice, but the grand and opulent connotations of the word 'Imperial' have clearly not been lost on the developers who have been selling the area to wealthy residents since the 1980s.

Kensal Green

Kensal Green today has an unfortunate and perhaps unfair reputation as being one of the most run-down areas of London. It has been much the same story since the early nineteenth century, when a sharp rise in population saw farmland redeveloped for residential use, creating cramped early housing estates.

The increase in population meant a greater demand for gas, and a works was built here in 1844 by the Western Gas Light Company. The location of the site was next to the Grand Union Canal, which allowed for coal to be delivered by barge.

But choosing a stretch of canal that ran through Kensal Green was also a strategic decision for other reasons. The Western knew there was huge potential in supplying gas to the wealthy districts of Notting Hill, Westbourne Park and Ladbroke Grove, and even further afield towards Kensington. Attracting business in these areas was of particular importance as the company was look-

Gasholder frames at the former Kensal Green Gasworks.

Former gasworks offices on Ladbroke Grove.

ing to compete with the Gas Light & Coke Company and Imperial Gas Light & Coke Company by offering a higher quality, purer gas. It would come at a premium, which only those with a higher income would be able to afford.

To actually build a gasworks in any of these areas would hardly have endeared the Western to their affluent potential customers, and so a location on the fringes was chosen instead, and Kensal Green provided the perfect fit.

It was a strategy that paid off, and the company soon had a firm grip on this area of north-west London – much to the frustration of their rivals. The Western would later annoy the competition even further by extending their reach into central London, reaching as far as Marylebone and even Bloomsbury. As with so many other smaller gas companies, however, it was to be the Gas Light that ended up benefiting most from the Western's success, as amalgamation came in 1872. The Gas Light invested heavily at their newly acquired works at Kensal Green, updating the machinery in 1889 and again in 1906. It was then fully rebuilt and modernised in on a grander scale than the original, although capacity was in fact scaled down during the same period.

The works at one time included its own basin on the Grand Union Canal, but this has been filled in. On the opposite side of the canal from the works is Kensal Green Cemetery, which was one of seven large burial grounds built as part of a project started in 1832 in response to London's churchyards being too overcrowded to handle the amount of deaths from decades of population growth. Famous names laid to rest at Kensal Green include Marc and Isambard Kingdom Brunel, Charles Babbage, Marcus Garvey, Harold Pinter and Charles Rattigan.

Gas production has long since ended at Kensal Green, and almost everything was demolished after closure. There are a few previous relics nevertheless, including an office block from the 1920s that can be seen on Ladbroke Grove, next to the entrance to a branch of Sainsbury's. There are also two surviving gasholders. These are best viewed from inside the grounds of the cemetery, or from along the canal towpath. Walking along the canal also allows you to see traces of where the gasworks' own basin used to branch off from the main waterway.

The site isn't technically abandoned, and the land is still currently owned by British Gas, who use it for storage. Part of the land is an open area of industrial wasteland that seems ripe for development, and various different plans have been proposed in recent years. It also featured in early drafts of the plan for Crossrail, but these were later changed, denying Kensal Green the chance of some urgently needed rejuvenation.

Vauxhall

It is clear from these pages that London has many well preserved gasholder frames. The most iconic of them all, however, are located in Kennington. They are seen by millions of television viewers every year, each time a test cricket match is played next door at The Oval.

They belong to what was once Vauxhall Gasworks, opened in 1833 by the London Gas Light Company. It was a business with a bad reputation, earned through some of the ruthless tactics they used to generate customers, including price undercutting and theft of other gas company's mains. They were nevertheless thriving by the 1930s, and built the site at Vauxhall to supply to various local districts, most of which were classed as being in Surrey until 1889 (today Vauxhall sits in the London Borough of Lambeth).

By 1860 the works had become too small, with little room to expand thanks to the development of London's main-line railways. In 1838 the London & South Western Railway (L&SWR) opened the first section of its ambitious new route that would eventually run from Southampton to a terminus station at Nine Elms. Although ideally placed for freight trains taking goods to and from the Thames, Nine Elms station was badly placed for passengers. Plans were drawn up therefore to extend the line to a more suitable location, where a new terminus station could be built.

The extended route connected Nine Elms to Lambeth, crossing through the heart of Vauxhall. The new train line meant that several streets and buildings had to be cleared to make way, and the new terminus station opened in 1848 as Waterloo Bridge Station, renamed in 1886 as simply Waterloo.

The construction of the railway had changed Vauxhall significantly, including the gasworks, which had been reduced in area by the new route. It was for this reason that expansion was out of the question, and instead the London Gas Light Company decided to close the works in 1865 in favour of building a large new site at Nine Elms (see the later section in this chapter).

This of course meant that in the space of less than twenty years, the industrial landscape of both Nine Elms and Vauxhall had witnessed a curious crossing of fortunes. The opening of Waterloo resulted in the closure of the station at Nine Elms for passenger services (a goods yard was maintained). Vauxhall had therefore lost its gasworks but gained a new railway station close by, whereas Nine Elms had lost their railway station but gained a gasworks.

The London Gas Light Company was absorbed into the Gas Light & Coke Company in 1883, and the Vauxhall works were later demolished. Nothing remains today, but they were located close to the Thames at Vauxhall Bridge, near to what is now Vauxhall Underground and National Rail station.

The gasholders were located some distance away, likely due to the lack of room on sight because of the railway. Three distinctive gasholder frames still exist near to the cricket ground, and they are best seen from Vauxhall Street and Montford Place. Vauxhall Street also includes an art and events space respectfully known as the Gasworks, housed in a building that was possibly once part of the works.

It is easy to imagine that the gasholders were a necessary eyesore that the owners of The Oval had to contend with when building the cricket stadium, but in actual fact the holders were built in 1853, several years after the stadium opened.

It was built in 1845 by the Surrey County Cricket Club, on a plot of land under the jurisdiction of the Duchy of Cornwall, which at the time was overseen by Thomas Pemberton Leigh. In addition to its use for cricket, the station also played host to one of the first international football matches, and the first ever FA Cup final.

Several hand drawn illustrations of the international football match clearly show the gasholders in the background, as the stadium was much lower until it was redeveloped in 2005. The inclusion of the gasholders in the illustration underlines just how closely they are connected with the history of The Oval, and it is this that has been used as a key argument against their demolition for redevelopment. Cricket commentators also often refer to the side of the stadium with the gasholders visible as the 'Gasworks End'.

Former gasworks buildings converted for commercial use.

Note that some historical accounts make reference to the gasholders at The Oval as being part of a different gasworks owned by the Phoenix Gas Company, which was later absorbed by the South Metropolitan Gas Company. The exact nature of the Phoenix Company's involvement with this site is unclear, as conflicting historical reports also list them as owning the main Vauxhall site belonging to the London Gas Light Company.

Rotherhithe

The south-east London district of Rotherhithe is a place with a rich maritime history. From being the launch site of the *Mayflower* to its heyday as part of one of the busiest docks in London, its legacy lives on in the names of its streets and landmarks. Surrey Docks, Canada Water (formerly Canada Docks), Jamaica Road and Quebec Way are just some of the many places where the echoes of the past can still be heard, and even Rotherhithe itself is said to have come from the Anglo-Saxon words for 'sailor' and 'wharf'.

One of the largest docks in the area was Greenland Dock, which by the beginning of the nineteenth century had become known for its many timber yards. It was built by the Commercial Dock Company, who later merged with other companies to create Surrey Commercial Docks. Today Rotherhithe is part of the London Borough of Southwark, but it was once classed as being part of Surrey. This explains why the docks were named as such.

It was also part of the name of the company that built a works in the area in 1851. Known as Rotherhithe Gasworks, they were opened by the Surrey Gas Consumers Company, founded in 1849. The company had already taken over a small gasworks in Deptford and was now looking to expand to Rotherhithe. It was an area that until now had only a basic gas supply for street lighting, delivered by mains pipe from a different gasworks in Deptford that was owned by fierce rivals the Phoenix Gas Company. Their coverage of the Rotherhithe area included the provision of lighting for the opening of the famed Thames Tunnel in 1843, but the Surrey Company was now poised to claim the district themselves.

Although still a relatively small company in comparison to the majors, they were able to take charge of Rotherhithe, and later expanded their reach into Bermondsey and other parts of Southwark. It began to make them the envy of several larger companies that had been looking at ways in which to penetrate the market south of the river.

The result was a long-fought and dirty battle between the Gas Light & Coke Company and the South Metropolitan Gas Company. They both wanted to absorb the Surrey Company into their empire, and it would prove to be one of the rare occasions where the Gas Light didn't come out on top. The South Metropolitan officially took ownership of the Rotherhithe gasworks and every other Surrey Company site in 1879.

Concerns over the danger of having a gasworks so close to the timber yards of the docks meant that the works had been built a short distance away from the river. The new owners were keen to construct a new dedicated wharf, however, and managed to bypass the concerns about fire hazards by simply purchasing several of the factories and timber yards that stood between the gasworks and the Thames. This allowed for coal to be delivered far more easily.

Gas production ended at Rotherhithe in 1959, after more than a century of operation. The site was demolished in the years that followed, but there are still some traces left to see today. They include a distinctive gasholder frame on Salter Road, which had been added to the original site in 1935. In recent years its height has come in handy for mobile phone companies, and part of the frame now has an aerial mast attached to it.

The abandoned coal jetty also still survives in the water, and can be seen along the Thames Path, parallel to Rotherhithe Street. West of the gasworks site can be found Neptune Street and Moodkee Street. They include residential buildings known as Neptune House, Murdoch House and Clegg House, and all three were originally built for employees of the Rotherhithe works. Two are named in honour of early gas pioneers: William Murdoch and Samuel Clegg.

Note that in some historical accounts, the Surrey Gas Consumers Company is referred to as the Surry Gaslight Consumers Association.

Solitary gasholder at the former Rotherhithe Gasworks.

Stratford

In summer 2012 the eyes of the world were focused on east London, as Stratford hosted the Olympic Games to much success. It was the culmination of years of planning that had seen the local area redeveloped beyond all recognition, making vast improvements for a new generation.

Lost somewhere in the long process was much of the area's rich industrial heritage, with several abandoned factories, warehouses and wharves in and around the Lea Valley either being demolished or cleaned up. Walk just a few minutes away from the Olympic Park though and it doesn't take long to find the remains of the industries they tried to hide during the games.

Some have been preserved thanks to being granted official listed status, including Three Mills at Bromley-by-Bow, and Abbey Mills Pumping Station located next to Abbey Creek (see chapter 5). Other sites have proven to be too expensive to remove, and have instead been scaled down for other use. It is this latter category into which fall the remains of Stratford Gasworks.

They were built in 1846 by a new company known as the West Ham Gas Company (indeed some historical records list the works as having been named West Ham Gasworks). It was a firm created by local businessmen in order to work in conjunction with the council authority on a plan to provide gas lighting to the surrounding area. The small fact that some of the company directors also worked for the council was the cause of some controversy, leading to allegations that contracts were only awarded because of the conflict of interest.

The new works opened regardless, and were enlarged several times over the years in order to reflect the company's expansion into Essex. Areas within their catchment included Loughton, Woodford and Chigwell.

Although not located directly next to the canal, the works were close enough for coal to be delivered by barge. This was then transported across the works via a private narrow gauge railway. A coal depot also once stood close by in what is now Stratford Market Rail Depot, and it is likely therefore that the gasworks were later supplied with coal directly from the main-line railways.

The West Ham Gas Company was absorbed into the Gas Light & Coke Company in 1910, who upgraded the site with modern equipment to help increase capacity. The gasworks continued to manufacture for several more decades, although the exact year of closure is unclear. The abandoned site was demolished some years after, but one of the gasholders remains in place today and is used for storage.

As an active gasholder with a telescopic mechanism, it is not always possible to see the holder from outside of the site. But when in its highest position, it can easily be seen from Rick Roberts Way, near to Abbey Road DLR station and the Stratford Market depot.

Remains of a former gasholder at Stratford Gasworks, West Ham.

Original gasworks wall at the former Stratford site.

Parts of the original gasworks wall can also be seen along the section of Abbey Lane that leads off from Rick Roberts Way, and it is possible that the small strip of houses along here were built for workers at the site. The small facility to which the remaining gasholder belongs today is named Abbey Lane Gas Depot.

Nine Elms

As described in the earlier section about Battersea Power Station, Nine Elms is an area south of the river with a rich industrial heritage. The railways had arrived and transformed the area beyond recognition, bringing even more factories and warehouses to Battersea and Wandsworth. The only thing missing from this polluted stretch of Thames waterside was a gasworks, but it finally came in 1858 thanks to a firm known as the London Gas Light Company.

Established in 1833, the company had until this point been manufacturing gas at their modest works in Vauxhall. Having overcome an earlier reputation of being a company that produced poor quality gas, they had begun to achieve much success by the 1850s and had now outgrown the size of the older site.

The new works at Nine Elms were large for their time, and were expanded several times to increase output. Similar to the many large power stations that would be built along the banks of the Thames in the century that followed the opening of the works, Nine Elms was one of the first sites to realise the benefits that being located next to the river offered for coal delivery. Most gasworks at the time had to rely on barges, but Nine Elms was able to save money by having coal delivered in bulk quantity by a fleet of large collier ships.

Even the coal shed where it was stored was somewhat ground-breaking, as one of the earliest buildings in Britain to have a roof constructed of corrugated iron. It was an innovation developed by engineer Henry Palmer, who designed it for use in the London Docks. It soon became one of the most widely used building materials in the construction of industrial buildings, and is still used around the world today.

The highly volatile gas production process meant that explosions at gasworks in the nineteenth century were common. Most caused minimal damage, but Nine Elms witnessed a devastating blast in October 1865 that left several men dead. The tragedy occurred when one of the two gasholders caught fire and exploded, shooting flames high into the sky. The other gasholder caught fire soon after, although fortunately it did not explode. Six men died instantly in the blast, with a further three dying later in hospital from their injuries. Countless other workers were also badly hurt, with injuries ranging from small cuts and wounds to horrendous burns. Among those that lost their lives was a boy named John Dwyer, aged just 16.

It was an explosion so immense that the ground shook with such force that it could be felt for miles around, and many houses were damaged. The exact cause is unknown, and the deaths were ruled to be accidental. The accepted theory is that the blast was the result of a leak in the gas line that may have been the result of unintentional damage caused by staff working in the meter house building. This was located close to the gasholders and was completely destroyed in the explosion. It is said that the spark needed to ignite the leaking gas may have been from one of the meter house workers breaking the strict rules on smoking. Ironically, one of the few surviving workers on duty in the meter house at the time of the blast is said to have left the works to go and buy tobacco.

The extensive damage was repaired remarkably quickly. Rival companies also assisted the London Gas Light Company by helping to maintain supply to their customers during the reconstruction work. A new, bigger gasholder had already been planned before the explosion, and it was completed and put into use earlier than expected to make up for the loss of the destroyed one.

The company was the subject of many takeover deals in the latter part of the century, but the directors tended to refuse most offers. It was finally amalgamated with a larger company in 1883, however, and once again it was the Gas Light & Coke Company. As ever, the new owners invested heavily in modernising the works, but the entire site sustained heavy damage during the Second World War.

The solution was to completely rebuild the gasworks in 1952, this time on a bigger scale than before. Space was at a premium, however, and some of the newly installed gasholders had to be located off the main site on a plot of land next to Battersea Power Station. But just twenty years after the remodelling work, Nine Elms was closed; another victim of the fast decline in gas production caused by the change to use of natural gas.

It was later demolished and today nothing remains of the gasworks. It is still a largely industrial stretch of the river, albeit far cleaner industries than before. The works were located on what is now a huge Royal Mail sorting office known as the South London Mail Centre. Also close to the former gasworks site is the New Covent Garden Market, opened in 1974 when it was moved here from Covent Garden itself.

But while the actual works may have disappeared without trace, the three gasholders built next to the power station are still intact, and can easily be seen from Prince of Wales Drive. Two are typical of the metal frames seen elsewhere, but the third is a different type altogether. Known as a dry seal holder, they were developed in the 1920s as an alternative to telescopic holders. Mainly seen on the continent, not many exist in the UK, and the only other former gasworks in London with a holder of this type is at Southall. The small site appears to still be in use for storage, and is now known simply as the Battersea Holder Station.

It is unclear if the holders will feature in any future plans to redevelop Battersea Power Station. Most schemes put forward in the past fifteen years have included an option for a new Tube station to be built on a section of the power station site, as part of a new extension to the Northern Line. The gasholders are located precariously with railway lines on either side, and may well have to be demolished to make way for the new Tube line.

Harrow and Pinner

There were gasworks in Harrow and Pinner, both of which now fall under the London Borough of Harrow, but were previously part of the Municipal Borough of Middlesex. They are towns that owe their development to the coming of the railways in the late nineteenth century, in particular the Metropolitan Railway, who opened a station at Harrow in 1880 as part of an extension from Baker Street (the station was renamed Harrow-on-the-Hill in 1894). A station was opened at Pinner in 1885 on an extension that headed yet further away from London, deep into the Buckinghamshire countryside.

Towns like Harrow and Pinner, often nothing more than extended villages surrounded by farmland, were now connected to the heart of London. New houses were built at a frantic pace, as London's workforce was encouraged to move to the new suburbs and commute into town. It was a successful concept that brought thousands of new residents to these areas, creating even bigger new towns that within a few short decades would be classed as being extended parts of the metropolis. With the people came a demand for gas, and several small but established companies were ready to supply it to them.

In fact, Harrow has had its own gasworks since 1855. It was the vision of a small business owner named John Chapman, who formed the Stanmore Gas Company after gaining the support of local residents and council representatives, all of whom were keen to bring gas lighting to Harrow's streets. Chapman built a modest works on Northolt Road, near to where South Harrow Tube station would be built forty-eight years later. Production at the works meant that existing oil lamps along the high street could be replaced by gas lights, which made a huge impression on the locals, and helped the town to grow.

The success of the company led to the works being rebuilt and expanded in 1872 by a new firm, the Harrow District Gas Company. Realising that competing for customers from the same works was hardly a recipe for long-term financial success, the two firms combined in 1894 to form one concern, which was duly renamed the Harrow and Stanmore Gas Company. By now the Metropolitan Railway extensions had arrived, and the new company was able to meet the growing demand each time a new housing estate was completed.

By the mid-1920s the continued success of the works at Harrow made them an attractive proposition for several larger companies looking to expand their reach, and in 1924 the Harrow and Stanmore Gas Company was absorbed by the Brentford Gas Company (see the following section on Southall Gasworks). Two years later the Brentford Company was itself amalgamated with the Gas Light & Coke Company, giving the larger company command over huge parts of north-west London.

In addition to bringing an influx of new consumers to Harrow, the Metropolitan Railway also helped the gasworks on a practical level. The works relied on coal delivery along the Grand Union Canal by barge, yet they were actually located nowhere near the water. This meant the company had to pay huge costs not only for delivery by barge, but then for the coal to be taken from canal side to their works by horse and cart.

A more cost-effective solution came with the building of Harrow-on-the-Hill station, as it included an extensive goods yard where coal could be delivered. There was still the logistical issue of having to move the coal from the station to the works, but it was a much shorter and therefore cheaper journey than from the canal.

The situation improved again at the turn of the century when the railway was extended towards Uxbridge on a route built largely on a new viaduct that ran very close to the gasworks site. Seizing the opportunity to have coal delivered by train directly to the works, the company applied successfully for a spur line to be built from the new branch line. The Harrow works were reconstructed to make way for a new set of sidings, and the spur opened for the first time in 1910. The works would go on to operate for another forty-four years before it was finally closed in 1954.

Pinner also benefited from the prosperity that the railway brought with it. A small gasworks had been built there in 1880 by the Pinner Gas Company to provide street lighting in the immediate area, and later to significant parts of Middlesex. It was to be a fairly short-lived company, however. After fifty years of operation, the works were amalgamated into the Gas Light & Coke Company. It was deemed to be too small a site to be maintained, and the decision was made to close the works in 1931. The area that Pinner supplied was instead connected to the mains from Beckton, miles away on the other side of London.

The gasworks at Harrow were demolished in 1986 and no trace remains today. The land has been redeveloped, and the site of the works now plays host to a branch of Waitrose. Accounts from local residents suggest that gas holders could be seen from Stanley Road and Brember Road, an area now home to an industrial estate.

What does remain, however, is a fragment of the spur line that can be seen from Roxeth Green Avenue. Although the line that once branched off to the gasworks was demolished two years after closure in 1956, a short section of the viaduct has been retained, and can clearly be seen branching off from the viaduct still used today by Piccadilly line trains travelling between South Harrow and Rayners Lane. It makes for an odd-looking structure, marooned up high and left to the elements. It has been claimed that demolition would potentially weaken the main viaduct, and therefore this last fragment of Harrow Gasworks looks destined to be preserved at least for now. Nothing remains of the works at Pinner, which were formerly located on Eastcote Road.

Southall

A basic gasworks operation still exists today at Southall in west London, but it is a different facility to the original works opened here in 1870. Its origins lie with the formation of the Brentford Gas Company in 1821, who constructed an earlier gasworks in the centre of Brentford, close to the Thames.

The company failed to run at a profit for several years, with many consumers bemoaning the poor quality of the gas being produced. It was the result of inefficient machinery and a lack of available funds to update it. The financial struggle continued until the late 1830s when a significant rise in population across many west London suburbs caused a new demand for gas.

The Brentford Company was now able to secure contracts to supply to parts of Ealing, Acton, Barnes, Hounslow and further afield. More success came in 1856 when the company was also given powers to supply gas for daytime use, primarily for cooking and heating.

The move towards mass use of gas in the home was a trend that was beginning to benefit several of the early gas companies, but in particular those such as the Brentford Company, who had focused their resources on supplying to public streets instead of merely the commercial sector.

They were finally now a successful company, and plans were put into place to extend their operation in order to keep meeting the growing demand. The problem was that the works at Brentford were located in what was by now a built-up area – another result of the population boost – close to the Brentford High Street of today. This made expansion impossible, and instead permission was granted in 1868 for a brand new works to be built in Southall.

They opened for business two years later in 1870, and would run successfully for more than 100 years, employing thousands of local people as the company continued to grow towards the end of the nineteenth century and into the next.

The success of the Brentford inevitably attracted the attention of the all-consuming Gas Light & Coke Company, and they amalgamated in 1926. With yet another large gasworks now added to their empire, the Gas Light

Distinctive gasholder at the former Southall Gasworks.

expanded the site at Southall in 1930, including the addition of a huge new gasholder. It was the largest holder built at the works, adding further capacity to the collection of original ones installed by the Brentford Company between 1880 and 1892.

The production of gas created several by-products that could be used for various different applications. It was therefore common for gasworks to earn additional revenue by selling chemicals to industry. Southall was one such site, and dedicated a portion of the facility to its Southall Product Works. It was a side of the business keen to be exploited by the Gas Light when they took over, and they expanded the operation to increase the output of products it could sell. With their works at Beckton also having its own chemicals division, the company was able to cover London from both sides, and the products work at Southall ran successfully until being scaled down in 1935.

Changes in the gas industry and decline in production led to Southall Gasworks being decommissioned in 1973. The site was retained for gas storage only, and continues to be used in that capacity today. What remains is a scaled-down version of the original, with most of the older buildings now demolished.

This has left a large area of wasteland that has caused considerable controversy in recent years, as local authorities clash with City Hall over how best to redevelop the site. For now, much of the land is used as satellite parking for Heathrow Airport.

Two gasholders can still be seen today, including a distinctive blue structure similar to the one close to Battersea Power Station that once belonged to Nine Elms Gasworks. These can be spotted from the entrance to the site on Beaconsfield Road, or from the other side of the works on Spencer Street. Nothing remains of the original works at Brentford.

Bow Common

With many parts of east London today suffering from poverty and a lack of investment, it is an area that at first glance seems like a million miles from the City of London. They are, in fact, inherently linked by history and geography, the latter via the A11. It is a road which starts in Aldgate, before passing through Whitechapel, Mile End and Bow as it heads out of London towards Norfolk. It is part of a route that has been used for trade in and out of London for centuries, and in the nineteenth century it became one of the most sought-after gas main lines.

Aldgate lies just inside the boundaries of the City of London, and it was the Square Mile that all emerging gas companies wanted to control – especially the Gas Light & Coke Company. It was an area they already had some control over, but this was threatened in 1816 by the emergence of a new company known as the City of London Gas Light & Coke Company. Official agreements were made between the two companies to share supply across the City without crossing each other's borders, but it was a compromise that still caused unease amongst the shareholders of the Gas Light.

It was a situation that became even more troublesome for the industry leader when another new firm entered the market. It was known as the Great Central Gas Consumers Company, and was created under somewhat pretentious motives. The principles of a 'consumers' company ruled that they should be driven by a desire to provide gas at a low and fair price for their consumers, and that the profits made by directors should be capped.

On the surface they were honest and noble aspirations. But it was often questionable whether such companies were genuinely fighting for the consumer, or simply using these principles as a way of undercutting their competitors in order to make greater long-term profit.

Regardless, their arrival and alleged intention to aim for the City of London territory sent even bigger shockwaves through the board room at the Gas Light than the plans of the City of London Gas Light & Coke Company had done. To add insult to injury, one of the key board members of the Great Central was a man named Alexander Angus Croll. He was a former employee of the Gas Light at their works in Brick Lane, and had been a pioneer in developing new advancements regarding the gas purification process. Croll had left the company in 1848 after becoming frustrated by the slow take-up of his new ideas. He claimed that his innovations could lower the cost of gas manufacture, which then meant that gas could be sold to the consumer at a lower price.

It was a debate that Croll and other financers of the new consumer company used to generate support from local people, and a special Act of Parliament was introduced that gave them the powers needed to start supplying gas.

A small works was built actually in the City of London, believed to have been located on Wood Street, near to London Wall. It was a strategic choice, and now made it clear beyond all doubt to other companies that the Great Central was indeed gunning for power over the City.

In a rare act of desperation, the Gas Light responded to the threat of the new company by lowering their own gas prices. It was a move that angered the City of London Company, who claimed that lowering prices would add weight to the Great Central's accusations that major companies had until now been selling gas at inflated prices. It also meant that two companies in their territory were now offering gas at a lower price than they were.

With three companies now fighting for control over the City, talks of amalgamation quickly began to circulate. The Gas Light and the City of London Company both made offers for the Great Central, but each was rejected.

With confirmation that they were now the envy of London's gas industry, they held firm and set about making plans to widen their reach towards east London. This was facilitated by the opening of Bow Common Gasworks, near Mile End, which placed the Great Central in the middle of a key territory dominated by the Commercial Gas Light & Coke Company. The new works at Bow Common were also built very close to the Commercial Company's own major works at Stepney.

The prospect of having a new competitor on their doorstep caused grave concern for the Commercial Company, but their fears were soon eased when it transpired that the Great Central had been refused permission to supply outside of the City of London.

It was positive news for the Commercial Company at first, but then gave rise to a new concern, albeit one being channelled through the company by the Gas Light. With permission for supply in east London denied, the Great Central dedicated their resources to expanding the control they had over the City. Plans were made for a series of long-distance gas mains to be laid from Bow Common all the way to the City of London, following the route of what later became the A11.

Part of the route required the new mains pipe to cross the canal, very close to one of the Commercial Company's existing mains. Supported by the Gas Light and City of London Company – both keen to stop the Great Central from gaining more power in the City – the Commercial Company set about stopping the new main from crossing the water. It led to the series of events described in the previous section as the Battle of Bow Bridge.

By 1870 the company had let the works at Bow Common become dated and in need of repair. It was the result of years' worth of failure in the years that had followed its battles with the major companies. The company's downfall was ultimately due to their founding principle: low-price gas being sold to a limited amount of consumers will inevitably lead to low profits.

The company had also been hit with several misfortunes that cost them dearly, including large fines they were forced to pay after negligence had caused a fire at their original works in the City. They were also the victim of financial fraud by a former employee, who managed to flee the country with the money he stole before he could be convicted.

The company was in such financial debt that the directors were forced to use their own personal money to keep it afloat. It was just one of many reasons why the company was compelled to accept the offer of being absorbed by the Gas Light in 1870. It was a deal that instantly relieved the Great Central's directors from their financial burden, and finally allowed the Gas Light to take full control of the City of London.

The remaining years at Bow Common were far less dramatic. It was expanded and ran successfully, until the nationalisation of the industry saw its owners disappear.

Since then the works have been used for storage, and two of its existing gasholders can be seen from Bow Common Lane, near to the corner of Ackroyd Drive. The well-preserved but abandoned office buildings can also be seen to the left of what was once the entrance to the works (note that access into the old site is not open to the public, and if the telescopic gasholders are in their lowest position it may not be possible to see them from Bow Common Lane).

It is unclear exactly where the Battle of Bow Bridge took place, but logic would suggest that it was probably where the A11 crosses the canal along Mile

Entrance to the former gasworks site.

End Road, close to Mile End Park. Nothing remains of the Great Central's other works near London Wall.

Note that in some historical accounts, Bow Common Gasworks are referred to as Bow Common Lane Gasworks.

Silvertown

Silvertown is an area of east London that has been home to many different industries over the past 150 years. It is even named after an industrialist, one Samuel Winkworth Silver, who built a large rubber manufacturers here in the mid-nineteenth century. Today, its cramped residential streets are flanked on one side by one of modern London's biggest success stories, and one of its most historic factories on the other. The former is London City Airport, opened in 1987 on what was once the Royal Albert Docks. The latter is the Tate & Lyle sugar refinery.

The famous food company has been based in Silvertown ever since 1877, when Henry Tate built a refinery. Tate's company was in fierce competition with a rival firm owned by Abram Lyle, but the two later amalgamated to become one entity.

The combined company expanded their operation in the first decade of the twentieth century, purchasing a plot of land that had recently become available after the closure of a gasworks that supplied to hundreds of consumers close by and further afield.

It was known as Silvertown Gasworks, and was opened in 1864 by a company known as the Victoria Docks Gas Company. It was founded in 1858 and quickly established itself with plans to expand. First, however, it needed a gasworks, and the solution was to buy an existing one near to Silvertown, in North Woolwich. This was acquired by amalgamation with the works' owner, the North Woolwich Gas Company.

One of the reasons behind the company's eagerness to expand was the ambitions of its chairman, George Parker Bidder. One of the most intriguing, albeit little-remembered, figures of Victorian London, Bidder was a child prodigy who was forced to travel the country and demonstrate his advanced mathematical prowess. As an adult he became an influential engineer, pioneering many innovations that improved London's new railways. By 1850, he had become chief engineer of the Victoria Docks, and oversaw their construction. It is this association with the company that led to his involvement with the new gas firm.

The works at North Woolwich were too small, and instead the company built their own new site in Silvertown. Its close proximity to the Gas Light & Coke Company's new works at Beckton made it the subject of amalgamation talks almost from day one. The inevitable decision was made in 1870, a mere six years after its opening.

As the owner of both sites, concerns about competition between Silvertown and Beckton were no longer an issue for the Gas Light. The closeness also meant that there was little sense investing in upgrading Silvertown, and gas manufacture was gradually phased out at the site. It was instead used primarily as a testing ground for new innovations, while Beckton picked up the slack by supplying to those who formerly purchased their gas from Silvertown.

The works closed in 1908, clearing the path for Tate & Lyle to move in just months afterwards. Despite the new owners having no connection with the gas industry, Silvertown's gasholders were kept intact for a number of years after closure, before finally being demolished in 1914.

Nothing remains today, but the gasworks were located roughly near to where North Woolwich Road meets Factory Road.

Beckton

If the power stations at Battersea and Lots Road were the crowning achievement of London's electrical supply history, then the former gasworks at Beckton were without doubt their equivalent in the gas industry. In its prime it was the largest works in Europe, spanning a huge plot of land in East Ham. Today, Beckton has established itself as a suburb, but it wouldn't even exist if it wasn't for the gasworks that once stood nearby.

Yet despite its vast size and the legacy it has left, the most surprising thing about Beckton Gasworks is that it has disappeared almost completely without a trace. This makes them not only London's most spectacular gasworks, but also the site perhaps most deserving of the title 'lost'.

They were built by the Gas Light & Coke Company between 1868 and 1870 as their new flagship headquarters. Decades of amalgamation with smaller companies had given the company several existing works across London in every direction. They had not, however, built a new gasworks of their own since their original, ill-fated sites in Shoreditch.

It was now time for them to create a works that would match the size of their success, which meant that the new site had to be huge. It was part of an aggressive new phase in the company's history, with the aim being to build a works so powerful that it would be capable of supplying gas to almost all areas of London.

It was also a reaction to the success of the Imperial Gas Light & Coke Company, who had in fact become more successful than the Gas Light by the mid-1860s (amalgamation with the Imperial Company would come later, in 1876). The Imperial Company had managed to build their success on the principle that a small number of very large gasworks were more effective than several smaller ones. It was an idea that would also later be advanced by Sebastian de Ferranti in the electrical industry. The Gas Light now decided to adopt a similar plan, and the powers required to build the proposed new works were granted in 1868.

Construction began soon after, on a huge plot of land in what was then considered part of East Ham. By 1870 the works were ready to start manufacturing gas. They had been spearheaded by Simon Adams Beck, one of the most influential men ever to work for the Gas Light. Beck rose through the ranks to become deputy governor, and later governor. With the works complete, the only thing missing prior to opening was a name. After much discussion, the company's board members agreed to show recognition to Beck's hard work by naming the new site after him. Thus the new site was christened as Beck's Town, later shortened to Beckton.

The size of the works meant that the company no longer now needed their other gasworks at Westminster, Brick Lane and Curtain Road, and all three were closed down a short time after Beckton opened. A new main was laid from East Ham all the way to Westminster, allowing the Gas Light to continue supplying to central London, despite having closed down their original works.

Within a few short years the Beckton site was supplying gas to much of London, and was expanded several times to meet the growing demand. The demand grew so much, in fact, that the Gas Light were unable to expand Beckton fast enough, and several amalgamated gasworks that were originally earmarked for closure were now forced to continue manufacturing gas.

Remains of the coaling jetty at the huge former gasworks at Beckton. (Image used courtesy of Paul Talling)

Regardless of how long it took to complete each new expansion at Beckton, each one made it larger and larger, and by 1900 the works resembled an entire town. Per square metre, it was larger than the City of London. Considering the Gas Light had previously fought so hard to take control of the City, the fact that their new gasworks was bigger than the Square Mile gives some sense of just how big the Beckton site was.

A chemical product works was also added to the site in 1879, allowing the company to sell the by-products of gas production, such as coke and sulphur.

Beckton had opened at a time when it was becoming increasingly cost-effective to use railways to receive coal and export chemicals. It was common, therefore, for gasworks to start constructing their own railway sidings which could be connected to existing main-line routes. As befits a works built to such a scale, Beckton not only had its own sidings, but also its own dedicated branch of the East County and Thames Junction Railway (EC&TJR), including a station inside the grounds of the works.

Primarily used for freight, the so-called Beckton Branch was also later used for passenger services that took workers to and from the works on a route that ran between Stratford and North Woolwich. Passenger services continued to Beckton until 1940, but the railway line would still be used for transporting chemicals from the works until as late as 1970.

Nationalisation of the gas industry led to the decline in gas production at Beckton, and it was closed for good in 1969. Its story was not about to end just yet, however.

As with several of the disused power stations discussed earlier, abandoned gasworks like Beckton were ideal filming locations for TV and film directors looking for a gritty landscape. Several productions in the 1960s and 1970s included scenes shot at Beckton, but one film in particular would have a major impact on the future of the site.

Visionary film-maker Stanley Kubrick had moved to Britain in the 1960s after directing the film *Dr Strangelove,* and later developed a fear of flying that kept him here for much of his life afterwards. As a result, most of his films made in the '70s and '80s were shot in the UK, including the 1987 Vietnam War epic *Full Metal Jacket.*

It was decided that the soon-to-be-demolished Beckton Gasworks would make the perfect location for a mock-up of the Vietnamese city of Hue. In order to make it look authentically like a city ravaged by war, several of the remaining buildings at Beckton were partially demolished or blown up. The sense of realism was successfully portrayed on film, but it left what remained of Beckton in ruins. The fragments of buildings still standing after filming had also been decorated with Vietnamese writing, which meant that years after the film crew had left, the gasworks site still resembled a bombed-out foreign city.

The cast and crew of the film had complained about the poor air quality and contamination at Beckton, which included traces of many different toxins as a result of its time as an active gas and chemical works. The entire site was later cleared, including the remains of the film set.

Part of the abandoned site also included a large slag heap, nicknamed after closure as the 'Beckton Alps'. The name was a reference to its significant height, but the alpine connection was justified even further in 1989 when parts of it were used as a dry ski slope which officially opened for business with a ceremony attended by none other than Diana, Princess of Wales.

Today Beckton Gasworks has almost completely gone. Most of the land has been redeveloped as an industrial estate and a retail park known as Gallions Reach Shopping Park. Opposite the shops, along Armada Way, one remaining gasholder frame stands proud along with a few small buildings close by. Aerial views of the works today reveal traces of at least six more gasholders, and an extensive network of blue pipes seen from street level suggests that some of these may still be used for storage.

Walking further along Armada Way reveals that large parts of the former works site still remain as wasteland, and have now become a haven for wildlife. Although not accessible to the public, the partially demolished remains

The surviving gasholder frame at Beckton Gasworks.

of Beckton's huge coaling jetty are still standing on the banks of the Thames, with hundreds of metal columns currently being left to rust.

Closed in 2001, some of the abandoned remains of the ski slope at Beckton Alps can be spotted from the appropriately named Alpine Way. The disused railway line has also mostly disappeared, but some parts of the route are now used by DLR trains travelling between Gallions Reach and Beckton stations. The original Beckton railway station was located on what is now a roundabout where Woolwich Manor Way meets Winsor Terrace (the street named in honour of the Gas Light's creator).

In addition to providing the entire area with its name, one former employee at Beckton Gasworks also helped shape the future of British politics. It was during his time as a worker at Beckton that Will Thorne created the National Union of Gasworkers and General Labourers in 1889. He became an influential trade unionist, and the organisation he began at Beckton became the foundation stone for the GMB.

Bromley-by-Bow

The east London district of Bromley-by-Bow has seen much redevelopment in recent years, with many derelict buildings cleared or converted for new use. But scratch below the surface and you'll find traces of an industrial heritage that has shaped the landscape for over 200 years.

Centred around Bow Locks, where Bow Creek meets the Lee Navigation, it was once an area known for its various mills. Although these have now mostly disappeared, their legacy lives on in names like Three Mills and Abbey

Mills. There are also still several miles of canal side littered with abandoned factories and wharves. Dominating the skyline above everything else, however, is a collection of gasholders, seven in total, that mark what was formerly Bromley-by-Bow Gasworks.

Constructed on the site of an old rocket factory, the works were owned by the Imperial Gas Light & Coke Company, who planned for them to be built on a scale large enough to compete with the Gas Light's mega works in East Ham. The decision to locate the works in Bromley-by-Bow was also the result of pressure from the government to encourage each major London gas company to start building their large works away from densely populated areas, in order to make living conditions better for residents across the city.

The deal to acquire the site was also conveniently assisted by the fact that the former rocket factory was owned by the Imperial's very own Sir William Congreve (see Pancras Gasworks section).

Work started in 1870 and within three years production of gas had begun, with coal delivered by barge via a new dock built as part of the works. The dock was named St Leonard's, a reference to an ancient priory destroyed during the dissolution of the monasteries from 1536 to 1541. It was later renamed Cody Dock, which is currently being redeveloped.

It was a site built to a large scale that cost the company hundreds of thousands of pounds, but it was to be money that the Imperial would struggle to ever get back. The immediate surrounding area was already adequately supplied with gas, and therefore there was little custom left for the new works to aim for. It was a huge commercial failure as a result, and one of the key reasons why the company accepted an offer to become amalgamated with the Gas Light & Coke Company in 1876, just a few short years after the works opened.

Several of the remaining gasholders at Bromley-by-Bow.

With their major gasworks at Beckton relatively close by, the Gas Light already had a large operation in east London. It was decided therefore that the Bromley-by-Bow site would remain untouched, with no plans in place for expansion or modernisation, except for the construction of some of the gasholders seen today. The works could have benefited financially by reducing the costs associated with coal delivery and storage if they had built a connection with the main railway line close by at West Ham. It was, however, an investment that the Gas Light was not willing to make.

The works continued to remain as built until they were finally closed a century later in 1976, by which point the site had been downgraded for use as storage only.

One thing that was added by the Gas Light was a peaceful memorial garden, built adjacent to the gasworks in a small area of woodland. Similar to the plaques at the Imperial's other works at Fulham, the park included a memorial for Bromley-by-Bow employees killed in the First and Second World Wars.

The Gas Light's own reputation as the most important firm in London's gas supply history is also represented in the park by a statue of Sir Corbet Woodall, which was moved here from Beckton after its closure. Woodall was governor of the company from 1906–16, a tenure that saw him help the company weather a difficult period. He was later knighted, and his position of governor only came to an end due to his untimely death in 1916.

The works were partly demolished after closure, including one of the gasholders, but today there is still much to be seen. The showpiece is undoubtedly the remaining gasholder frames, each of which has a unique style to its design. It is rare for such a large collection of frames to exist intact, and their close proximity to each other makes them appear as though they are interconnected. They are best observed from Twelvetrees Crescent, which allows you to get almost within touching distance of them.

A more dramatic view can be achieved from the canal towpath near to Three Mill Lane, where the gasholder frames can be seen in context with the canal on either side, and a handsome railway bridge that takes District and Hammersmith & City line trains across the water as they travel between Bromley-by-Bow and West Ham. The gasholders have been Grade II listed since 1984, which should be enough protection to see them preserved for many more years.

The memorial garden also still survives, and can be found hidden away in a clearing of trees on the opposite side of Twelvetrees Crescent to the gasholders. It was only open to employees of the gasworks in its heyday, but it is now open for everyone to enjoy. In addition to the war memorials and Woodall statue, the garden also includes an original gas light that burns constantly,

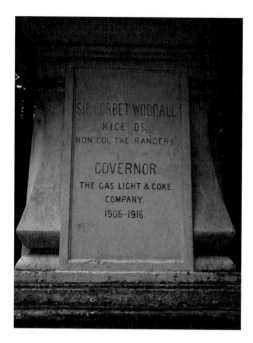

Statue located in the memorial park at the former Bromley-by-Bow Gasworks.

providing a poignant reminder of what was once here.

Also close to the memorial park is a building that was once used as offices for the gasworks. It was later redeveloped as the London Gas Museum, but this closed in 1989. The artefacts once on display are now held by the National Gas Museum in Leicester.

Note that some historical accounts list the works at Bromley-by-Bow as being named simply Bromley Gasworks. The confusion likely stems from a different gasworks built in 1864, located in the south-east London district of Bromley. This site is now known as Homesdale Road Gasworks, and has its own gasometers which can easily be seen from Liddon Road.

East Greenwich

The Greenwich Peninsula is today thriving as a new centre of entertainment for London, with the O₂ Arena, Emirates Air Line, hundreds of new apartments and much more bringing life back to what was a barren industrial landscape for much of the twentieth century.

It is a far cry from what it was like in 1800. Back then the peninsula was simple marshland known collectively as Greenwich Marshes (it was also sometimes referred to as Bugsby's Marshes, which is a name believed to be derived from a pirate whose body may have been displayed here along with several others after their execution, as a warning to those passing by). It was used by farmers as grazing land for livestock, and was a haven for wildlife.

It was a peaceful scene but one that wouldn't last much longer. Its location, flanked by the Thames on all sides, placed it at the heart of the developing dockyards and shipbuilding industry. By 1870 the peninsula was surrounded by the Millwall Docks on the Isle of Dogs to the west, the East India Docks to the north, and the Victoria and Albert Docks to the east. It was a significant change that had seen the marshlands disappear and be replaced by hundreds of warehouses and wharves.

Businesses that came to the peninsula included many involved with the manufacture of components for shipbuilding, plus steelworks, chemical plants and an arms factory. With hundreds of workers employed by the various new companies, the peninsula also developed into a significant residential zone, with whole communities emerging as more firms continued to set up shop.

Having water along each side also made the peninsula a prime location for a gasworks, and in 1880 permission was granted to the South Metropolitan Gas Company to build a large new site here. It was to be the last major gasworks that London would ever see, and was built on a huge scale that was considered to be the South Metropolitan's answer to the Gas Light & Coke Company's works at Beckton.

The new works were the vision of Sir George Livesey, who had by now risen to the rank of director at the South Metropolitan, having spent several years as their chief engineer. Livesey was part of a family deeply connected with the history of gas supply in London. First came Thomas Livesey, who was appointed as deputy governor of the Gas Light in 1815. It was a position he held until 1840. His nephew, also named Thomas, was also employed by the company, and by 1839 was working in the role of chief clerk at their early gasworks in Brick Lane.

The younger Thomas then left the Gas Light to become secretary of the South Metropolitan – much to the disgust of his uncle. He played a key role in setting up the new company, and it was he who would later bring his own son George into its employment. It was far from a case of simple nepotism, however, as George Livesey quickly established himself as a man with a clear sense of exactly how a successful gasworks should be built and run.

His first experience with the company came when he was appointed to work at their gasworks in Old Kent Road, which had begun producing gas some years earlier in 1834. It was a works plagued by financial failure and industrial accidents, but it nevertheless helped the South Metropolitan to become established as a new force in the industry.

Decades later, with George Livesey now installed as director, and work at East Greenwich about to begin, the company was thriving, largely thanks to the work of his father Thomas. George's plan for East Greenwich was that it needed to be constructed on a massive scale if it was to realistically compete with Beckton.

Work took several years to complete, and by its opening in 1889 the new gasworks had come to dominate the peninsula, taking up almost all of its central section. Several previous factory buildings were cleared during construction, and an extensive internal railway was built as part of the works, assisting with the delivery and movement of coal.

The works provided gas to many parts of south London, and proved successful enough to warrant expansion several times in the decades after

Above: East Greenwich Gasworks during its prime. (Image used courtesy of the Greenwich Heritage Centre)

Left: The former gasworks site in full operation. (Image used courtesy of the Greenwich Heritage Centre)

opening. The original layout included two huge gasholders that were reported to be the largest in Europe. One of them was badly damaged in 1917 as a result of one of the deadliest explosions in British history.

It occurred at a factory across the river from the East Greenwich works, close to the Royal Victoria and Albert Docks in Silvertown. Originally in the business of chemical production, the factory had been re-equipped for the manufacture of munitions in response to the shell crisis of 1915, where supplies of ammunition for the First World War had run short. The disaster occurred when several tons of TNT accidently exploded, killing a total of seventy-three people.

The blast was big enough to damage buildings a mile away, including the gasholder across the water at East Greenwich. No casualties were sustained at the gasworks, however, thanks to the efforts of many workers who helped contain the damage caused. Several were later recognised for their bravery by King George V.

Similar to Beckton, East Greenwich also earned extra revenue by opening a chemical works. But by the 1970s the major changes within the gas industry that had affected every other gasworks also took their toll here. It closed for production in 1976, and almost the entire site was demolished over the course of several years, including the giant gasholder that had been damaged by the explosion at Silvertown.

The major redevelopment of the Greenwich Peninsula means that the area is now unrecognisable from its gasworks days, with several new roads and buildings now constructed on where it once stood. One of the original two gasholder frames does still exist, however. Its presence looms large over everything new that has arrived here in the past twenty years, and it can be found standing proud on Millennium Way.

The memory of the great George Livesey was ridiculed somewhat by a bizarre story that circulated in the British tabloid press in the late 1990s, during construction of the Millennium Dome. It was claimed that the redevelopment of the old gasworks site had stirred the ghost of Livesey, and that several workmen at the Dome site had heard his 'laughter'. Considering the failure of the Dome project from start to finish, it was a story that handed the newspapers a convenient anecdote on which to hang the widespread criticism of the new building.

The Dome has proved to be far more successful since reopening in 2007 as the O₂ Arena, but even the new owners have not been able to shake off the rumours of the Livesey ghost. In a press story even more ludicrous than the first, it was reported in 2012 that pop star Lady Gaga had asked for supernatural experts to be called to the arena so that they could banish 'bad energy' while she prepared to play a concert. It was quickly tied to the myth of

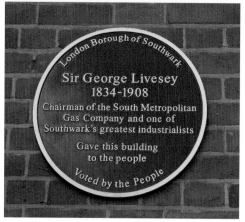

Above: The only surviving gasholder from the huge former gasworks on the Greenwich Peninsula.

Left: George Livesey's plaque on Old Kent Road.

Livesey's angry spirit, although unsurprisingly, no trace of anything sinister was actually discovered.

Livesey is remembered in a more appropriate light near to the gasworks at Old Kent Road, where he worked from a young age. At number 682 Old Kent Road is an abandoned building with the words 'Camberwell Public Library No 1' written in the stonework. It was built in 1890 and donated to the people of the area by Livesey himself. It served in its function as a library until closure in 1966, but was then reopened in 1974 as the Livesey Museum.

Remaining gasholders at Old Kent Road Gasworks.

The opening ceremony was presided over by Sir John Betjeman, and several years later a statue of Livesey was moved from outside the Old Kent Road gasworks and relocated here.

The museum was sadly closed down in 2008 due to funding cuts by the London Borough of Southwark. The statue of Livesey was removed after closure, and its whereabouts are currently unknown. The museum building includes a blue plaque dedicated to its founder. The gasworks at Old Kent Road are located close by, where several of its original gasholder frames are still intact. These can best be seen from Devon Street and Sandgate Street.

Livesey is also remembered in Sydenham at the Livesey Memorial Hall. Outside stands the Livesey Hall War Memorial, erected in honour of workers from the South Suburban Gas Company who died during the two world wars. The company was founded in 1853, and later absorbed several smaller gasworks that gave them control over parts of Crystal Palace, Bromley, Dartford and Tonbridge. Their flagship gasworks were located close to the memorial hall, and its gasholder frames can be seen from Perry Hill. The hall and war memorial at Sydenham, as well as the former museum and statue at Old Kent Road, are all now Grade II listed structures. George Livesey died in 1908 and is buried in Nunhead Cemetery.

HARNESSING THE MIGHTY THAMES

Although steam played an important role in the generation of electricity across London's growing network of power stations, it was often a supporting role at best. Electricity was clearly the way forward, and steam was merely one part of the process, as opposed to its main component.

Away from the major plants, however, steam-powered engines still had much to offer as a method of power generation in their own right, although they delivered it in a way that was already starting to feel nostalgic in the mid-1800s.

Today steam power is often looked back upon in a romanticised fashion, along with everything else that was so good about the Industrial Revolution that started in Britain. It is clear that electrical networks like the National Grid have enriched the modern world. But plugging into the mains and flipping a switch lacks any of the raw mechanical thrills that only a steam engine can generate.

In Victorian London, steam engines were being used across the city primarily as a means of powering different types of pumping station. There are several of these still in existence, most of which have been preserved by dedicated enthusiasts, and are explored later.

HYDRAULIC POWER STATIONS AND THE THAMES SUBWAY

Steam was also used as the driving force behind a hydraulic power network that covered large parts of central London, encompassing several power stations. The basic principle of a hydraulic power system is the movement of a fluid substance at high pressure. London's original hydraulic power stations generated the pressure required by the use of steam-powered pumping stations. The energy the pumps generated could then be piped through mains directly into buildings, as a source of power for a range of different applications.

The heavy industry of the Victorian era and the decades that followed meant that hydraulic power was used first and foremost in London to work machinery. In particular, hydraulic power was in demand by the rapidly developing railways and the docks. The power was used in the goods yards of the major rail companies as a way of lifting heavy wagons. On the docks, it was being used to control the opening and closing of huge dock gates, and to power the cranes that loaded goods on to and off ships from around the world.

But it wasn't just industry that benefited from hydraulic power. It was also used in hotels, shops and offices for lifts, and even as a way of powering the stage and other mechanisms used in theatres and exhibition centres. The Royal Opera House, the Savoy Theatre and Earls Court were just some of the venues connected to the hydraulic system.

The dominating force in the city's hydraulic supply was the London Hydraulic Power Company (LHPC) and their vast network of pumping stations and mains. Although today hydraulic power can be achieved by applying pressure to various different liquids, in the nineteenth century the liquid in question was almost always water. This was particularly appropriate, of course, given that there was no better source of any fluid available in London than the mighty river that runs through it. It is hardly surprising, then, that the LHPC's network of pumping stations were located close to the Thames, with water taken directly from the river.

Some of the company's buildings have been demolished since the network closed down in the late 1970s, including their head offices in Pimlico (not to be confused with Western Pumping Station, which is still standing nearby and is covered later). Also long gone is the company's first power station near Blackfriars Bridge, known as Falcon Wharf, plus another station on the City Road Basin along the Regent's Canal. Several of the company's power stations do still exist however, and what follows is a look at what remains.

Wapping

Wapping Hydraulic Power Station can be found along Wapping Wall, close to some of the most historical buildings in London. It was opened by the

LHPC in 1890 and continued in operation until its closure in 1977. It was driven by steam on its opening, but was converted to use electrical turbines in later years.

London Hydraulic Power Company lettering.

It has been well preserved since closing, with the entire building still intact, including a large pump house and an accumulator tower. The site is Grade II listed and is now home to a restaurant and arts centre known as the Wapping Project. Its owners have maintained much of the original machinery, including several generators located inside the former pump house. It was the final LHPC site to close, and is said to be the last pumping station in England with its equipment still in situ.

The power station is located in a courtyard off the section of Wapping Wall close to Shadwell Basin. Its status as a functioning public building means that you are able to go inside and explore what remains.

Renforth

Another surviving LHPC power station is located in Rotherhithe, close to the river, on the opposite side to the station at Wapping. Known as Renforth Hydraulic Power Station, it was opened in 1904 and used to provide power for various different forms of waterside machinery. Closure came in the late 1970s with the rest of the network, but today the buildings have been successfully converted into apartments that boast incredible views towards Canary Wharf. It can be found on Renforth Street, close to the south entrance of the Rotherhithe Tunnel.

Chimney stack at the former Renforth Hydraulic Power Station.

Tower Subway

Although it is now mostly famed for the role it played in the development of London's underground railways, the history of the Tower Subway also involves the London Hydraulic Power Company, who purchased the tunnel in the 1920s.

Opened in 1868, it was the first to use a pioneering new type of tunnelling shield developed by James Henry Greathead and Peter W. Barlow. The new method of tunnel digging expanded on some of the basic principles as the system that had been used to build the more-famous Thames Tunnel twenty-five years earlier.

The new shield enabled the tunnelling process to become faster and safer, and its success served as the inspiration for the first network of deep-level underground railways anywhere in the world. The first section of the City & South London Railway (C&SLR) opened in 1890 between King William Street and Stockwell, and was built using the Greathead–Barlow method. Up until this point, the underground railways that would later form today's Tube network had been constructed using the cut-and-cover method. This involved the digging up of roads to make way for shallow trenches. Track would then be laid, and stations constructed inside the trenches. The new stations and track were then roofed over and the road re-built on top. It is for this reason that many stations still in use today have platforms only a few steps down from street level, in particular those on the Circle line.

But while cut-and-cover was a revolutionary feat of engineering, it was an awkward and expensive method of construction that often involved having to purchase property above ground in order for work to progress. The deep-level tunnels built by the C&SLR were cheaper to make, and allowed greater scope for expansion. As discussed earlier, the line also pioneered electric trains on the underground network, therefore also putting an end to the problem of how to get rid of smoke from the steam trains that worked the cut-and-cover lines.

The tunnelling shield developed for the Tower Subway is essentially the same system that has been used to build rail tunnels in London and around the world since. Yet while the legacy of its construction method still lives on in engineering terms, the subway failed as a business venture when it opened to the public. The plan on opening in 1870 was for the tunnel to be used to carry passengers under the Thames via steam-operated cable cars that each seated twelve people. It was a somewhat unpleasant and claustrophobic ride however, and low usage led to its closure just three months after opening.

It was then adapted into a free-to-use pedestrian foot tunnel which at first proved popular. Unfortunately, it later became a haven for thieves and prostitutes, a fate also shared by the Thames Tunnel in the years prior to its conversion for use by the railway. The fatal blow for the Tower Subway came with the opening of Tower Bridge close by in 1894. Also free to use, it provided a far more pleasant way to cross the river, and the subway closed for good in 1898.

By this time the London Hydraulic Power Company's network was well established, but getting power across the river had always been a major challenge towards achieving a city-wide network. Pipes were already installed along Waterloo, Vauxhall and Southwark bridges, but greater capacity was needed.

The prospect of being able to utilise a disused tunnel under the Thames was therefore an attractive one, and an Act of Parliament allowed the company to buy the abandoned Tower Subway for use as part of the network. A new set of mains pipes was installed, which enabled hydraulic power to be transferred to both sides of the river with ease, between the power stations at Wapping and Renforth.

The company demolished the original entrance shafts to the tunnel and replaced the one on the north side of the Thames with an elegant brick-work structure. This still stands today, hidden away next to the ticket offices and food stalls for the Tower of London next door. It is well preserved, and proudly displays the words 'London Hydraulic Power Company' across the top. It hasn't been used for hydraulic power since the company shut down in 1977, but is now used for general water mains and telecommunication cables.

Small traces of the LHPC can also be found across London if you look hard enough, including several cast-iron box covers that were used to conceal the valves through which the mains from the system were connected to indi-vidual buildings. A few examples of these are located along Clink Street. The company also provided back-up power for Tower Bridge, and its own small museum features much of the original machinery.

ACCUMULATOR TOWERS

Heavy users of hydraulic power networks, such as railways and dockyards, were able to store the power supplied to them with the aid of an accumulator tower. Invented by William Armstrong, these tall brick structures housed huge tanks where water could be retained. When large amounts of power were required, massive weights above the water tanks would be used to apply force, therefore exerting enough pressure upon the retained water to generate the energy needed to power machinery.

Royal Mint

Once a familiar site across London, most accumulator towers have been demol-ished, but there are several fine examples still standing today. One such building can be found close to Tower Bridge, near the junction of Mansell Street and Royal Mint Street. The lettering on the side of the building is all but faded, but once proudly displayed the words 'London Midland & Scottish Railway City Goods Station and Bonded Store', which reveals its original ownership and use.

Abandoned Royal Mint accumulator tower.

Its origins lie in the history of the London & Blackwall Railway, which ran from Minories to Blackwall and the London Docks, along the viaduct next to the accumulator tower. Minories railway station was located close by, opened in 1840. It closed around fourteen years later when the opening of Fenchurch Street station made it largely redundant. The station and its surrounding area were then redeveloped as a goods yard between 1858–62, known as Royal Mint Street Depot. The accumulator tower is said to have been built in 1894, and would have been used to power lifts in the goods yard for the movement of wagons. The complex closed in the 1940s and was later demolished, leaving only the accumulator tower intact. Today the DLR runs along the viaduct, and standing at the far end of the platform at Tower Gateway station (built on the site of the original Minories station) allows you to get a great close-up view of the accumulator tower structure. The London Midland & Scottish Railway lettering on the side refers to one of the amalgamated rail companies with an interest in the site long after the original London & Blackwall Railway had disappeared.

Blackwall accumulator tower, now housing a wine merchant.

Blackwall

Further along the original London & Blackwall Railway route is Blackwall DLR station, close to which is another building with an accumulator tower, located on Blackwall Way. Before redevelopment in the 1980s, the surrounding area was dominated by Poplar Dock, including a complicated network of railway goods sheds with a hydraulic power network. The accumulator tower was built in 1877, and was part of a larger collection of buildings. The rest have been demolished but the tower and pump house building still remain. It is now a wine warehouse, meaning it is possible to go inside and take a look at parts of the structure.

Camden

Another accumulator tower related to the railways can be found in north London, hidden behind a collection of modern apartments on Gloucester Avenue in Camden. Although much of the site has been demolished, the surviving tower was once part of Camden Goods Depot, a huge complex originally opened in 1839, and subsequently involved with the North London Railway (NLR), London and Birmingham Railway (L&BR), and Midland

Railway. Several buildings on the site have been preserved and refurbished for new use, including the Grade II listed Roundhouse and the Stables Market, while others have been lost.

The accumulator tower is said to have been built by the London & North Western Railway (successor to the L&BR), and would have served a similar function to the Royal Mint and Blackwall towers. A pump house was added later, although all traces of this seem to have disappeared.

Limehouse

Perhaps the most striking of all London's remaining accumulator towers is the one located in Limehouse, near to Mill Place. Its distinctive shape is in stark contrast to the other towers, but it operated in exactly the same way. It appears, therefore, that the different shape was purely aesthetic, and another example of how in Victorian London even the most functional and unglamorous industrial structures could be ornate and stylish.

It was built in 1869 and was used to power a hydraulic network in the Limehouse Basin, used to operate cranes and lock gates. The basin opened in 1820 after being built by the Regent's Canal Company. The use of hydraulic power was introduced in the 1850s, including components designed by William George Armstrong. The current tower was built as a replacement to an earlier one, and continued to be used until the 1920s. It was restored in the 1980s and is now Grade II listed.

It is occasionally opened to the public, but some of the best views of the tower can be seen by riding the DLR between Limehouse and Westferry. Note that despite being so close to the original London & Blackwall Railway route, the accumulator tower was not connected with the railway.

SEWAGE PUMPING STATIONS

Pumping stations were also used for a wide range of other processes across London. Although not strictly related to the production of electricity or gas, they were driven by the power of steam, and there are several surviving structures worthy of inclusion here.

The basic principle of a standard pumping station is the same as that of a hydraulic power station. The difference is that instead of being used as a method of moving pressurised water over long distances, common pumping stations were used as a way of helping gravity to move water from one place to another. In London, it was usually either to or from the Thames.

The use of pumping stations helped to play a crucial role in helping London solve an undesirable but necessary problem that had plagued the city for centuries: how to get rid of its sewage. Up until the 1850s, London had a long history of dumping its human waste directly into its rivers. One of the most infamous was the River Fleet, the water from which was further polluted by the blood and offal from slaughtered livestock at Smithfield Market, as it made its way towards the Thames. The Fleet is one of several lost rivers in London, buried underground in order for redevelopment above.

The Thames itself was also used as a dumping ground for waste, with primitive sewers and drains running from public buildings directly to the mighty river. By the early 1850s, tens of thousands of people were dead from cholera, and the growing issue of human waste disposal came to a head with the so-called Great Stink of 1858, when the smell across central London became so unbearable that action was finally taken by the powers that be.

Parliament and the Metropolitan Board of Works had to act fast, and in the end it was the genius of engineer Joseph Bazalgette that solved the problem. His idea was simple but effective. It involved the construction of underground sewers that could carry water from homes and buildings out of London, so that it could then be deposited into the Thames Estuary, and therefore far away from the central part of the river used for drinking water.

Much has been made elsewhere of the ornate Victorian design work of these original brick sewers, most of which have only ever been seen by a select few people, and of how Bazalgette's system also reclaimed parts of the Thames to form the Victoria, Albert and Chelsea Embankments.

Less often remembered, however, is that the sewer network also included several pumping stations, each one beautifully designed to a degree of quality that transcended their actual use. They were needed to help sewage water flow through areas where gravity alone was insufficient.

Abbey Mills

Drawing comparisons with Bankside Power Station being designed as a 'cathedral of power', Abbey Mills Pumping Station came to be known as the 'cathedral of sewage'. Although perhaps not the most flattering name when first heard, it was in fact a compliment to this extraordinary building and its grand design.

The name Abbey Mills is said to originate from a watermill that once existed on the same site. Its location is one of several points through which water flows along Bazalgette's original sewer route as it heads towards the Northern Outfall Sewer near Beckton (its counterpart is the Southern Outfall Sewer, which controls flow underneath boroughs in south London).

Illustration of Abbey Mills Pumping Station on its opening. (© Science Museum/Science & Society Picture Library)

Construction was completed in 1868, boasting a Byzantine design that was worked on by renowned Victorian architect Charles Henry Drive, Edmund Cooper and Bazalgette himself. The building has a Grade II listing but still includes working pumps that can be used as back-up if needed.

The original steam pumps were replaced with electric motors during the station's working life in 1993, which would have rendered the huge two ornate chimneys redundant, had they not been demolished during the Second World War over fear that they would be the target of bombs. Their stonework stumps can still be seen today, and are also Grade II listed. A high-tech replacement pumping station also now stands on the site, built in 1997.

Abbey Mills Pumping Station is only accessible to the public on special occasions, such as Open House London weekend. It can, however, be seen from a public footpath known as the Greenway, located next to the former Stratford Gasworks site described earlier.

Crossness

Built on a similar grand scale to Abbey Mills, Crossness Pumping Station is located in south-east London, near to the Southern Outfall Sewer. Just like its sister station north of the river, Crossness was the work of Bazalgette and

Driver. It features the same exquisite design, octagonal internal structure and elaborate ironwork, which is again comparable to a church or temple.

It was completed in 1865 and opened by Edward VII (at this point still Prince of Wales). It was constructed using millions of bricks, inside of which were housed four engines. One was named after Edward, and two of the others named in honour of his parents, Queen Victoria and Prince Albert.

The building and its machinery faced years of dereliction after closure in the 1950s, but in recent years it has been fully restored, including one of the original steam pumps. Much of the building is now Grade I listed, and includes a museum that can be visited on special open days.

Western

Another pumping station from Joseph Bazalgette's original sewer network can be found near Chelsea Bridge on the north bank of the Thames, on the opposite side of the river to Battersea Power Station. Although not quite as elegant as Abbey Mills and Crossness, Western Pumping Station is nonetheless a stunning and well-preserved building, with Grade II listing.

The station was opened in 1857 and powered by four steam engines. These were later replaced in the 1930s by diesel engines. It isn't completely disused, as it still plays a functioning role in today's sewage system, with modern equipment housed in an adjacent building for emergency back-up.

The most striking thing about the pumping station is its tall brick chimney, which can be seen from miles around. Similar in appearance to an accumulator tower, and indeed perhaps used in a similar way, the chimney is also Grade II listed, and is today used as a ventilation shaft from the modern sewer below.

Western Pumping Station in Pimlico.

The disused remains of Deptford Pumping Station.

Deptford

There is yet another of Bazalgette's pumping stations based at Deptford Creek, close to Greenwich. It is not quite as stunning as the others, but boasts an Italianate design that still impresses. This makes the fact that it currently stands abandoned and forgotten all the more frustrating.

It opened in 1865 with a series of beam engines that elevated sewage as it followed on its route towards Crossness and the Southern Outfall Sewer. Despite its Grade II listing, it has been largely hidden away behind trees by current owner Thames Water. It can, however, be seen through railings along Norman Road.

Markfield and Walthamstow

There are two other surviving Grade II listed sewage pumping stations north of the river, both lovingly restored. Not strictly part of the original Bazalgette network, they still played a part in moving waste towards the Northern Outfall Sewer.

The Markfield Pump House is located in Tottenham and was opened in 1864. A beam engine was added to the site in 1886, and the plant operated well into the 1950s. Most of the machinery was removed after closure, but the beam engines were salvaged and have since been fully restored. The main pump house building has also been refurbished and is now home to a museum.

Walthamstow Pump House was opened in 1885 by the Walthamstow Urban District Council. Similar to the Markfield site, the former pump house building, known as Low Hall, is now a museum.

WATER SUPPLY PUMPING STATIONS

At the complete opposite end of the spectrum from sewage, London also had a small selection of pumping stations that were used for the process of providing drinking water to the suburbs. Pumping stations in this context would typically have been used to take water from a reservoir, which could then be piped into a public water supply system.

Kew Bridge

A surviving example of such a building can be found near the banks of the Thames at Kew. Built by the Grand Junction Waterworks Company between 1837 and 1871, Kew Bridge Pumping Station is now part of a steam museum that has preserved the original steam-powered beam engines at the site. The centrepiece of the museum is the Grand Junction 90 Engine; a huge beast of a machine whose size impressed Charles Dickens when he visited in 1850.

Dickens had come to Kew Bridge as research for an article titled 'The Troubled Water Question'. It was to be used as part of a campaign for a safer public water supply that Dickens had championed. The report, along with the work of John Snow, who discovered the link between dirty water and cholera, was influential in the passing of an Act of Parliament in 1852. The Act ruled that water taken from the polluted Thames had to be treated if it was to be used for drinking. The murky waters of the Thames were of course greatly improved in the decades that followed, thanks to the work of Bazalgette, but at the time, at least, it was still a major concern.

The engines at Kew Bridge were used to supply water to much of west London, taking it directly from the Thames. It was then filtered into reservoirs, included those located over seven miles away at Campden Hill, near Holland Park.

The site at Kew also includes an attractive brickwork standpipe tower that is reminiscent to the one at the Western Pumping Station. It houses several huge pipes that were used to protect the entire system by absorbing some of the water surges often generated by the pump engines.

As with most pumping stations across London, the Kew Bridge site was later converted to being operated by electric pumps, and finally closed in 1944.

Kempton Park

Another surviving water supply pumping station is located at Kempton Park in north London, and is part of the original Kempton Park Waterworks opened here in 1897 by the New River Company. It was a firm whose history can be traced back to 1613 and the construction of a man-made waterway known as the New River.

Running a distance of around 20 miles from Hertfordshire to Stoke Newington, it was built as a way of delivering clean water from the River Lea to north London, via a network of conduits, wells and aqueducts. The original route has been altered over the past 400 years, but sections can still be seen in parts of Islington, Enfield and Bowes Park.

The station at Kempton Park was used to pump water from the Thames so that it could be stored in reservoirs. Once treated, the water could then be pumped into the mains in order to supply directly to north London, in particular the area in and around Cricklewood.

The New River Company later came into the ownership of the Metropolitan Water Board, who expanded the site in the 1970s by building a new engine house. Two large steam engines were installed, capable of running non-stop, and powerful enough to pump millions of gallons of water every

Chimneys at Kempton Park.

day. The original engines were taken out of service in 1980 and replaced by electric pumps inside a new boiler house next to the original.

The two large engines have been preserved, however, including one that still works. Named after Metropolitan Water Board Chairman William Prescott, the working engine can be seen in action on special 'steaming' open days. The engine houses are beautifully designed, and the site also includes two immense brick towers, similar to the one at Kew Bridge.

It is worth noting that Kempton Park also had its own narrow-gauge railway that was used to carry coal from ships to the engine house. The coal was used by the huge boilers, which generated the steam needed to power the engines. Much of the railway has been lost, but a similar line has since been reconstructed as a feature of the museum at Kew Bridge.

OTHER LOST WATERWORKS

London also once included several other waterworks owned by many competing companies. Names included the London Bridge Waterworks Company, the Borough Waterworks Company, the Lambeth Waterworks Company and the Chelsea Waterworks Company. Most of the sites owned by such companies have since been lost, except for a few surviving reservoirs and parts still used in today's water supply network.

All of the companies listed above became part of the Metropolitan Water Board in 1903, as did the Grand Junction Waterworks Company. The board was a public body that lasted until 1974 when it was replaced by the Thames Water Authority. This was then privatised in 1989 as Thames Water.

CANAL PUMPING STATIONS

Pumping stations were also used on the canal network in order to top up water lost when a lock system had been operated. Little appears to be known about its use, but it is likely that a distinctive-looking pump house building next to St Pancras Lock was originally for this function. Taking into consideration that there is no other water source close by other than the canal itself, the pump house was possibly used for 'back pumping', a process where lost water from the highest level of a lock is pumped and replaced with water from a lower section.

An alternative theory is that this structure may have been part of the vast railway goods yard that once operated behind Kings Cross and St Pancras International, especially as its design is reminiscent of the railway-owned accumulator tower discussed earlier.

St Pancras Pump House.

Thames Tunnel Engine House

While most of the water supply and sewage pumping stations of the Victorian age were concerned with taking water from the Thames and then putting it back in later, the sheer power of the river itself was causing problems elsewhere, and it is even still a problem today.

The issue is a simple but potentially deadly one: if you are going to tunnel underneath the Thames, then how are you going to keep those billions of gallons of water above you from leaking in and causing a flood?

It was a problem first encountered by engineer Marc Isambard Brunel when he was designing the previously mentioned Thames Tunnel. As the world's first tunnel built under a major river, it is one of the greatest engineering works of all time, and one of the greatest among many achievements by Marc's son, Isambard Kingdom Brunel.

Fraught with major setbacks that included gas explosions, fires and crippling financial difficulties, the tunnel took eighteen years to complete. The biggest setback of all was the frequent flooding that plagued construction. The worst flood of all happened in 1828, causing the death of six men. Isambard Kingdom Brunel was himself inside the tunnel at the time of the flood and was almost killed.

BRUNEL'S ENGINE HOUSE

The tunnel shaft and pumping house for
Marc Brunel's tunnel was constructed
between 1825 and 1843. This was the
first thoroughfare under a navigable
river in the world.

Above: Thames Tunnel
Engine House.

Right: Plaque
commemorating the use
of the pumping station for
the Thames Tunnel.

It was an issue that Marc Brunel had anticipated from the very start, and his solution was to build a steam-powered pumping station that could be used to remove water from the tunnel. The pumps were installed inside the Engine House, which still stands today, and is located next to the original tunnel shaft in Rotherhithe.

Designed by Marc Brunel himself, it is a functional but attractive building made of brick, with an elegant tall chimney. The inside of the building is now home to the Brunel Museum, which details the tunnel's construction, its commercial failure when it was finally opened in 1842, and its later use by the railways.

The issue of leakage through London's tunnels under the Thames is still an issue today, and it is said to be one of the reasons why the Waterloo & City line on the London Underground is nicknamed 'The Drain'. It is also not uncommon to see traces of water when walking through the Greenwich foot tunnel.

Today the problem is controlled on the London Underground by the use of modern electric pumps that extract thousands of gallons of water from the tunnels every day. The threat comes not only from the Thames, but also from various other sources of groundwater across the city.

The sophisticated equipment used today is wisely hidden away from public view, but until 2012 a fine example of a Victorian pump house used for this very function could be seen in the Docklands on the site of the former Royal Albert Docks. A simple but well-designed brick building, it housed steam-powered equipment used to extract water from the Connaught Tunnel that runs below. The tunnel was built in 1878 to allow trains to run under the docks instead of over them via bridges.

Train services through the tunnel were abandoned in 2006 as part of the closure of a branch line that ran from Canning Town to North Woolwich. The tunnel was in the process of being refurbished at the time of writing, to be brought back into use as part of the Crossrail project. The upgrade includes new pumping equipment that requires a more significant building at ground level than the original brick structure. But all is not lost, as its rumoured that the old pump house will be rebuilt as a ticket office for the historic SS *Robin*, a restored steamship now located close by on part of the Royal Victoria Dock.

6

CLOSURE, ABANDONMENT & DERELICTION

There are many different external reasons why London has lost its power stations, gasworks and hydraulic power network, all of which will be discussed here. But on a very basic level, the decline was often simply a case of natural progression. Each was replaced by something more advanced and efficient, necessitated by demand as it continued to grow in parallel with the ever-expanding population of London and its many boroughs and suburbs. The following pages explore how each industry died, and the legacy they have left behind.

ELECTRICITY GOES NATIONAL

The decline of London's electric power stations was caused by a combination of economic, governmental and environmental factors. Some led to vast improvements to how electricity was supplied to Londoners, while others caused considerable controversy. But regardless of public opinion or the long-term consequences of how the industry changed, the outcome for the city's power stations is that all of them were closed down, abandoned and usually demolished without a trace.

The first significant change in Britain's electricity supply industry was in 1925, when the government responded to concerns about the lack of consistency across the numerous different supply companies and the power stations they owned. The problem was that competing companies often used different voltages, frequencies and delivery systems to generate and deliver electricity to the consumer. It made little difference to the public, but as an industry it was almost impossible to create a unified network that could connect the entire county to one system.

The man tasked with investigating the problem was William Douglas Weir. Born in Glasgow, Weir was one of the most influential men of the early twentieth century. He served as director for several major companies, including some related to the manufacture of gas, oil and chemicals. He was highly decorated in later life, with achievements that included two knighthoods, followed by peerage status as a baron. He would eventually become Viscount Weir, before his death in 1959.

Weir's contribution to the electricity industry included the founding principles of the National Grid. The findings of his report underlined how important it was for Britain to have one single electricity network, fully integrated via a small selection of large power stations. It was an idea that echoed the thoughts of Sebastian de Ferranti some years earlier. His justification for building a huge station at Deptford had been a belief that London needed just a few 'mega-stations' to cover the work previously undertaken by many smaller outfits. There had been great advancements in technology in the decades since Ferranti's pioneering work, however, particularly in relation to how current was delivered from power station to consumer.

The call from Weir to create a national network had been encouraged by impressive developments at a power station in Newcastle, named Neptune Bank. Opened in 1901, it was the first power station anywhere in the world to use a new system of high voltage delivery that meant electricity could now be transmitted across longer distances, and therefore more easily connected from one large city to the next. It was exactly the kind of solution that Weir argued should be rolled out across the country, and his findings resulted in a new Act of Parliament being passed in 1926, known as the Electricity Act.

In order to ensure that Weir's recommendations were executed, the Act led to the creation of a new organisation known as the Central Electricity Board (CEB). Using direct current (DC) as standard – again pioneered by Ferranti – the CEB introduced a grid system for electricity distribution across the country, connected by smaller, localised grids. After several years of testing and refinement, the new system was fully introduced in 1938, creating the National Grid that is still in use today. It was at this point that several of London's power stations began to close or become downgraded to being merely substations.

The large stations across London would also later suffer the effects of more change in the industry, but for now they helped play a major part in the new national distribution system. Several large sites were selected to form a ring around London, all connected together as one of the local grids. Among those chosen were the power stations at Battersea, Croydon, Fulham and Deptford, allowing firms such as the London Power Company to maintain their presence, despite now being part of a nationwide network.

In 1947, however, an updated version of the Electricity Act was passed which nationalised the entire British electrical supply industry. It meant that all privately owned and council-operated power stations were placed into the realm of public ownership. From municipal sites like Fulham to the LPC-owned Deptford and Battersea, all would now be operated by a new organisation known as the British Electricity Authority (BEA).

It wasn't of course just in London where the BEA now had control. In fact, across the whole of Britain, the new organisation now had authority over more than 500 former electricity companies, council departments and power stations.

It was run as one central organisation, with the country split into regional boards. The power stations in London fell under the jurisdiction of the London Electricity Board (LEB). As the branch in charge of the most demand-heavy region in Britain, the LEB was arguably the most powerful of all those under the control of the BEA, with no fewer than thirty council boroughs and twelve private companies absorbed.

This dramatic new scenario meant that after years of dominance, the London Power Company ceased to exist, as did every other private electricity firm in the city. Considering how aggressively companies such as the LPC had expanded during their heyday, and the sheer size of their mega power stations; the infrastructure, property, machinery and land now inherited by the BEA was significant to say the least. The BEA's remit also included the National Grid, and as a consequence it led to the demise of the Central Electricity Board.

Further changes occurred in 1954, when amendments to the power supply industry in Scotland made way for a new Act of Parliament. Under the BEA, Scottish supply was controlled by two of the regional boards, the South East Scotland Electricity Board and the South West Scotland Electricity Board (the northern part of the country was operated by a separate entity). The new Act handed control instead to the Scottish Office, which was a subdivision of the UK government. It was a change significant enough to see an end to the British Electricity Authority. Instead it was replaced by a new organisation known as the Central Electricity Authority.

Incredibly, just three years later, the CEA was also dissolved in favour of yet another organisation. It was the outcome of a new set of amendments to the Electricity Act in 1957, resulting in the creation of the Central Electricity Generating Board (CEGB). The new organisation absorbed everything that fell under the responsibility of each association that came before it, including all power stations and the National Grid. While the CEGB focused on actual generation and supply of electricity, the 1957 Act also included provision for a secondary body known as the Electricity Council. It was to serve as the policy maker for the government, and would later be influential in the next round of changes to the industry.

Irrespective of what each of the above organisations was called, they shared the same broad outlook for the electricity industry, and it was this that led to the closure and abandonment of London's remaining power stations. By the

1950s and '60s, London was a city that had become far more conscious of the heavy pollution created by its coal-fired power stations, especially those located so close to densely populated areas. The various new authorities born out of nationalisation were therefore under pressure to move power stations away from London.

There were also economic factors to consider, thanks to the ever-rising costs associated with transporting coal from the north into London (by now, collier ships had made way for mass rail freight). Further advancements in technology and the expansion of the National Grid meant that it was now possible for electricity to be carried over hundreds of miles via pylon and substations. Even for an area as demanding of power as London, enough bulk supply could now be delivered more cost-effectively from elsewhere.

The solution to the environmental and financial issues was therefore to construct new power stations well away from cities, and closer to where coal was produced. It led to mass closure of power stations across London from the early 1960s to the late '80s, including the great cathedrals of power such as Bankside, Deptford and Battersea. The only major exceptions were Lots Road and Greenwich, who continued to serve their purpose as the lifeblood of the London Underground.

It was an ironic development of Ferranti's original concept that a small number of large power stations were all that London needed to best serve its people. Modern stations were now so powerful elsewhere that London didn't require its own power stations at all.

The CEGB continued to manage Britain's electricity supply until 1990, when it was announced that the entire industry would be privatised. The CEGB therefore ceased to exist, and control of the country's supply was instead split between several different private companies. Most still exist today, albeit as subdivisions of larger multinational energy companies. The move towards privatisation was influenced by the Electricity Council, which itself was dissolved in 2001.

Privatisation meant that the power industry had essentially gone full circle, back to how it had evolved organically in the first place. It was too little, too late for London's power stations, although it is of course hard to deny that the city is better off without the smog and pollution they created.

In addition to Greenwich, London is today served by a small selection of power stations located on the outskirts of the city. One is situated at Barking, close to the original Barking A, B and C stations discussed earlier. Another is located in Enfield, on a site next to the former Brimsdown Power Station. There is also an existing station at Croydon, close to the original A and B site. Taylors Lane Power Station in Willesden is a relative newcomer. It was opened in 1979 during the Central Electricity Generating Board era, as a replacement to an earlier station built on the same site that operated from 1903 to 1972. All the stations mentioned above are operated by gas, a fuel source discussed in a later section.

GASWORKS KILLED OFF BY NATURAL GAS

The gas production industry fell into decline for many of the same reasons as the production of electricity, discussed earlier. Shifts in government, advancements in technology and concern over public health and environmental issues all contributed to the loss of gasworks across London and beyond. The entire gas production industry was also hit hard by the use of modern consumer appliances that ran on electricity, which were beginning to be used on a larger scale than ever before.

The first significant change came after the end of the Second World War. It was a major transitional period for Britain, with many people looking for change. The role played by Winston Churchill during the war had elevated him to iconic status, but post-war Britain was fast becoming a country that demanded social reform. It led to the shock victory of the Labour Party in 1945, led by newly appointed Prime Minister Clement Attlee.

One of the key principles of the new government was a belief that the country's workforce should see vast improvements with regard to rights, wage levels and health and safety legislation. The outcome was nationalisation on a significant scale, which moved several types of industry into public ownership. The list of heavy industries that were nationalised in the first few years of the new government included the railways, steel manufacture, the coal mining sector, electricity supply and the gas industry.

The new changes to the gas industry were formalised under the passing of the Gas Act of 1948. The inter-war years had been tough on the industry, with several smaller companies being declared bankrupt, having never fully recovered from the First World War. Each conflict had meant gas production at several gasworks had had to be stopped while they were converted for use as chemical production plants. Large parts of the workforce were also called into action, leaving most works understaffed.

The mass bombing campaigns suffered during the Second World War had also caused major damage to several of London's gasworks, meaning that hundreds of thousands of pounds would be needed for repairs once the conflict ended.

The gas industry was consequently in a state of disarray, and the new Attlee government commissioned a report to explore how it should be rebuilt. The finished document became known as the Heyworth Report, having been written by a committee headed by Geoffrey Heyworth. It concluded that in addition to post-war rebuilding, the industry would also benefit from being divided into regional sectors overseen by a central body. It was a concept that worked hand in hand with the idea of nationalisation, and it was these elements combined that formed the backbone of the 1948 Act.

The introduction of nationalisation meant that over 1,000 privately owned gas companies disappeared overnight as the industry was now under public control. Drawing comparisons with the electricity industry, Britain's gasworks were split into geographical zones, as per the findings of the Heyworth Report. Each section was known as an Area Gas Board, with twelve created in total. Somewhat differently to the report's suggestion, however, there was no central body as such. Instead each regional board was assigned its own designated chairman of the board. The majority of London's gasworks fell under the jurisdiction of the board known as the North Thames Gas Board, including those sites owned by the Gas Light & Coke Company and the Commercial Gas Light & Coke Company.

South of the River Thames, gasworks were designated as part of the South Eastern Gas Board, including the major site at East Greenwich. Although it held no responsibility over the Area Gas Boards, a Gas Council was also created to advise on policy.

After more than a century of operation, development, financial hardship, amalgamation and often brutal competition, all of London's gas companies no longer now existed, including the mighty Gas Light. The pioneering company had understandably been against nationalisation ever since it was first discussed. Considering that the state was inheriting the Gas Light's decades of experience, its staff and several large-scale gasworks, it was argued that significant financial compensation should be given to the company. It was to be a request that was largely ignored however.

The company was officially wound up in 1949, but its gasworks would still operate under their new ownership. They continued to manufacture town gas for public and commercial use, with many works now fully recovered from wartime damage. But in the early 1960s the gas production industry faced its biggest challenge yet. It was to be the final downfall for London's gasworks, and it was a threat that came from deep below the sea.

Following the example of natural gas fields discovered in America and other parts of the world, scientists discovered gas in the North Sea, off the coast of the Netherlands. Unlike town gas, which had to be physically manufactured via the process that occurred at a gasworks, natural gas existed deep below the ground.

As the name suggests, it occurred naturally, trapped inside thousands of years' worth of sedimentary rock formed by fossilised sea creatures and plant life. The gas itself was in fact a combination of different gaseous elements including carbon dioxide, nitrogen and methane.

It was first discovered in rock formations below land, often during the process of extracting oil. Finding gas occurring naturally under the sea was therefore a major scientific breakthrough, and research was conducted to see if other parts of the North Sea could also be farmed. Britain's scientists discovered huge reserves of gas off the coast of East Anglia, near Great Yarmouth and Lowestoft. Tests were carried out to see if it could be used for consumers, and it soon became clear that natural gas had huge advantages over town gas.

From an economic standpoint, extracting gas from the sea was far more cost-effective than the coal-dependent and intensive processes under which town gas was manufactured. Instead of hundreds of individual gasworks around the country, supply could now be controlled via a small number of sites, later known as gas terminals. These were to be built strategically, close to the source, with gas piped into them for purification. The processed gas could then be distributed to towns and cities using mains, including those already built by the defunct gas companies. As with polluting electric power stations being moved away from densely populated areas, gas terminals were built in isolated locations.

With natural gas now confirmed as the ideal replacement for town gas, the government launched an epic initiative in 1967 to convert the whole of Britain to the new form of gas supply. It was a scheme that involved every single household and commercial building in the country having its mains and appliances adapted. It took ten years and over £560 million to complete, killing off the gas manufacturing industry in the process. Town gas was now no longer needed; therefore there was no gas to manufacture. Gasworks began to disappear at great speed, including the mega-sized sites at Beckton, East Greenwich and Nine Elms.

The conversion to natural gas had taken place during another period of change for how the nationalised industry was operated. Under the Conservative government of Edward Heath, a new Gas Act was put into force in 1972, mid-way through the conversion process.

The new regulations absorbed the Area Gas Board structure, and instead created a new organisation with a name better suited to a nationalised industry: the British Gas Corporation. It was created under the scrutiny of the Secretary of State for Trade and Industry, who at the time was Peter Walker. It was later changed to being under the jurisdiction of the Secretary of State for Energy, and Walker was succeeded by Tony Benn after the Labour party regained power, led by Harold Wilson.

With the conversion to natural gas complete by 1977, the industry would again experience major change in the 1980s, under the controversial leadership of Margaret Thatcher. One of her most significant policy decisions was the selling off of national assets in order to raise cash that could offset deficits elsewhere. An amended version of the Gas Act in 1986 therefore led to the gas industry being privatised, and as a result the British Gas Corporation became the British Gas plc. When Labour regained power in 1997 under Tony Blair, the public limited company was wound up and instead split into different segments, most of which still exist today.

The complete decline in gas production is the reason why there are no longer any active gasworks in London. As discussed in the previous chapter, several former works still have active gasholders. These are used to store gas piped in from elsewhere, as a method of regulating how much gas is needed at any one time.

The sites are today managed by the National Grid Company, one of the groups into which the former British Gas plc was segmented into after 1997. Taking into consideration the fact that the National Grid Company also has major interests in the electrical supply, it is an interesting dynamic to have one company so heavily involved with two industries that spent much of the early twentieth century in competition with each other. It is now common for single companies to supply consumers with both utilities, a vision never realised in the heyday of London's separate gas and electricity companies. With many different firms selling energy to consumers, however, the competition for business is at least as fierce as it was a century ago.

The nearest gas terminal to London today is located on the Isle of Grain in Kent, opened during the first few years of the privatised British Gas plc era. Gas is piped here from countries in the North Sea area, primarily Belgium and Holland. New oil and gas fields continue to be discovered in the North Sea, although it has become an increasingly expensive procedure in recent years. It is for this reason that gas is also now delivered in liquid form to sites such as the one in Kent.

Another of the gas industry segments created during the changes of 1997 was Centrica, which operates today under the famous brand of British Gas. It is still the market leader, and the company that most people associate with the gas industry. It has direct links back to the Gas Light & Coke Company, meaning that even now, over 200 years after it was founded, the vision of Frederick Albert Winsor is still having an impact today.

HYDRAULIC POWER CAN'T TAKE THE PRESSURE

The fall of London's hydraulic power network and the company that owned it occurred not as the result of policy making or major advancements in technology and science. Instead it was simple commercial failure, as it was an industry unable to compete with electricity.

As early as the beginning of the twentieth century, the new-found success of the electric power stations – particularly the majors – hit the London Hydraulic Power Company hard. What their network offered was the end product of a brilliant engineering process, but it was flawed by its limited range of uses.

Hydraulic power had many applications for industry, but it was far harder to find ways to apply it successfully for domestic use. The new electricity power stations that emerged across London were not only able to generate power on a huge scale across long distances, they were also selling a commodity that appealed to both industry and the general public alike. For now, however, the LHPC's industrial contracts were enough to sustain their power stations.

But the company's fortunes were then impacted again, this time during the Second World War. The prolonged air strikes across London caused enough damage to necessitate large parts of the city having to be reconstructed. New buildings meant new foundations and changes to road layouts. Many of the company's water mains would therefore need to be re-routed.

Moving the mains would have been a costly business and hard to justify when considered against the low output the network was actually creating. The war had also caused damage to hundreds of commercial and industrial buildings, and as a result there was less demand than ever for hydraulic power.

The following decade, the company also lost many of its long-standing contracts as several London-based industries began to decline. Several railway goods yards were closed, and therefore no longer needed hydraulics to move wagons. The docks were also declining fast, and no longer required power to operate gates and cranes.

The historic London Hydraulic Power Company finally succumbed to defeat in 1977. As the only firm in London to operate such a network, it was to be the last time that hydraulic power would ever be used on such a large scale in London.

Most of the abandoned infrastructure was purchased by a company known as Mercury Communications. It was a new business created in the wake of the 1981 British Telecommunications Act, which encouraged new companies to challenge the dominance of British Telecom. Acquiring the assets of the LHPC gave Mercury a ready-made system of tunnels, mains and pipes they could use for wires. They were even able to use the Tower Subway to traverse that great divider of London's north and south, the Thames.

Today Mercury no longer exists, but parts of the former hydraulic system's mains are still used for communication and data transfer.

7

REDEVELOPMENT & NEW USES

HISTORIC POWER STATIONS AND GASWORKS SUCCESSFULLY REGENERATED

The many lost electrical power stations, gasworks, hydraulic pumping stations, accumulator towers and other such buildings across London often pose an awkward problem for developers.

From a cultural and academic perspective, the argument for keeping such buildings standing is a strong one. They are a major part of London's history, and deserve no less respect than any other structure that has played its part in the epic story of this great city.

The modern Londoner seems to be more aware than ever of the significance that important buildings have. One need look no further than the popularity of annual events like Open House weekend to see how much interest and intrigue there is to be found in London's past. Organisations like English Heritage have worked hard to help preserve many of the sites covered in this book, awarding several with official 'listed' status that ensures their protection for at least the foreseeable future.

For others, however, the visible remains of London's industrial past are often considered as nothing more than eyesores that contribute to urban decay. There are large parts of London – particularly in the East End – that seem to have been missed out of regeneration and redevelopment schemes that have modernised many other areas. In the same way that a large, crime-ridden housing estate can often hinder the rejuvenation of a residential neighbourhood, an abandoned and rusty old gasholder frame can sometimes have the same effect.

Another problem that commercial estate agents and property developers face is just how big the task of bringing an old industrial building back to life can be. It is often a lengthy process involving years of planning and new design work, before construction can even begin. When it is finally time to start, more issues have to be overcome. An old gasworks site, for example, involves decontamination of the land in order to remove hazardous chemicals and bring conditions back to a safe level.

It is reasons such as this that go some way to explaining why a building such as Battersea Power Station stands as empty and disused today as it did on closure in the early 1980s. It is clear that land is more precious than ever in London, with property prices dropping only slightly after the recession of recent years. It is therefore a source of great frustration for many – namely estate agents and local councils – that mega-sized pieces of land like Battersea, parts of the old Beckton Gasworks and various other places still stand empty.

There are, however, many sites included here that have been successfully redeveloped and reused. The key appears to be a happy medium approach that usually falls into one of two categories. The first approach is often to acknowledge a building's former use by preserving certain elements and using them as part of the redevelopment. The other is a more modern approach, where a building is repurposed with little connection to its past maintained, except for a few small original features or naming conventions. Both work well, and it is projects like these that are helping to ensure many old buildings can still be enjoyed by future generations. What follows is a look at some key examples at several sites explored earlier.

Bars, Restaurants and Galleries

Many of the former power stations and generating stations included here have been converted into restaurants and bars, including the Great Eastern Electric Light Station and Wapping Hydraulic Power Station. Buildings such as these are the latest in a long line of historical and former industrial buildings being brought back to life by the food industry. One example is the Oxo Tower, which itself was originally built as a small power station that was intended to be used by the Post Office. It is more widely known for its later use as a warehouse for a food company whose product range included Oxo cubes. Its top floor and roof top were opened as a restaurant in 1996. Another example can be found in Chelsea, where an upmarket restaurant has been in operation since 1997 in a Grade II listed building that was once the Bluebird Motor Garage.

But while restaurants such as these pay homage to their origins often in name only, several of the lesser remembered buildings included in this book have stayed more faithful to their past. The generating station in Shoreditch, for instance, now a bar/club, has been left almost exactly how it would have looked after the decommissioned boilers and generators had been removed. The ceiling also still includes an original pulley system device, and outside some signage has been maintained from the building's time as being owned by the railway.

Converted restaurant at Wapping Hydraulic Power Station, with boilers still *in situ*.

The former hydraulic power station at Wapping has gone even further by keeping much of the generating equipment in place inside the former turbine room, with diners in the restaurant seated around the old machinery.

The large steam engines, turbines and boilers that power stations relied on tended to take up lots of space. Once the machinery has been removed, what's left is a vast open space. This makes the buildings ideal for events and large-scale art installations, and the former power station at Wapping also runs regular gallery showcases. The industrial and dramatic backdrop of the machinery adds an extra level of appreciation to the sometimes rather pretentious art pieces. The Shoreditch Electric Light Station has also been converted into an event facility that makes full use of the space available.

The most famous use of a former power station as an art gallery is of course Tate Modern at Bankside. Its large size has provided the perfect location for displaying some of the biggest pieces of contemporary art ever created, most of which have been displayed in the massive former turbine hall. In the book *Power Into Art*, author Karl Sabbagh describes how the decision to reuse the abandoned Bankside Power Station was an important part of the planning process for what the gallery would become. Commissioning a new building specifically for the gallery would have run the risk of the architecture itself being a bigger draw than the actual artwork. Using an older structure instead allowed for the art to complement the building, and vice versa, by providing a link between the historic and the contemporary.

The desired effect appears to have worked, as visitors seem to appreciate both the artwork and the immense scale of the building that houses it. It is likely that many visitors aren't even aware it was once a power station, but this lack of historical knowledge is of little importance for such an incredible building having been given this fantastic new lease of life.

In 2012, a further connection to the Tate Modern's former history was created with the opening of the Tanks. They were located below the main power station buildings, and it was here where the gallons of oil needed to run Bankside were once stored. The additional space has allowed the Tate to showcase even more large installations. The fact that the artworks are now located in what was once something as filthy as an oil tank further underlines the irony of how those opposed to the power station in the late 1940s were concerned about the effect it would have on cultural London.

Commerce and Cleaner Industries

For power stations and gasworks that have mostly been demolished, the space left behind is usually snapped up by property developers and filled with new residential buildings. Some plots of land are more desirable than others, however, but it still hasn't stopped those less popular places from being monetised in some shape or form. Where the Barking Power Stations once stood is now an area of open space on an industrial estate. But once a week it turns into a hive of activity as it hosts a lively Sunday market. Customers arrive and park their cars next to a derelict building that most would not even realise was once part of a huge power station complex.

The space surrounding the surviving gasholders at Southall Gasworks now generates income as a car park advertised as being perfect for Heathrow. There's room for hundreds of cars, although in reality the airport is located some distance away. Original buildings at former gasworks like the ones at Fulham and Vauxhall have been reopened as commercial units for different businesses, while the Chiswick Power House has been converted into a high-end recording studio that has played host to a client list that includes Michael Jackson and the Black Eyed Peas.

New Homes for a New Generation

The former sites that have been cleared and redeveloped for housing have provided new homes for hundreds of families. The apartments constructed

Renforth Hydraulic Power Station, now converted into luxury apartments.

on land that was once Stepney Gasworks pay only subtle recognition to its history, but have brought affordable living to one of London's poorest boroughs, brightening up the area in the process. At the other end of the price spectrum, original buildings such as those at the former Renforth Hydraulic Power Station have been converted into luxury apartments worth millions.

At the time of writing an ambitious project is underway to build a complex of new apartments as part of the massive Kings Cross Central redevelopment. Once construction on the new homes is complete, three of the preserved gasholder frames from the former Pancras Gasworks will be rebuilt around each apartment block, cleverly combining the old with the new.

Major Landmarks Stuck in Development Hell

It is evident from the examples above that many former power stations and gasworks have found a new use since closure. But for some sites, redevelopment has been thwarted by bad planning, lack of funding or changes in the economy. The size of some of the projects means they are often deemed a failure on an epic scale, causing embarrassment for politicians and business leaders.

Take, for instance, the site of the disused East Greenwich Gasworks on the Greenwich Peninsula. It was earmarked for redevelopment in the mid-1990s as what would later become the Millennium Dome. It was a project started when Britain was under a Conservative government led by John Major, with Michael Heseltine spearheading an initiative to convert the abandoned peninsula for new commercial use. It was an area chosen over several others that were shortlisted, including land close to the Bromley-by-Bow Gasworks.

It was a time when the dawn of the new millennium was fast approaching, and when Tony Blair's Labour Party came to power in 1997 it was decided that the Dome would be one of their first major projects. Plans for this huge new structure to be used as the focal point for Britain's celebration of the new millennium became ever more ambitious, but it was soon clear that it was running enormously over budget.

Once opened, the Dome attracted millions of visitors, but still far fewer than expected or were needed to turn a profit, leading to huge financial problems that generated widespread criticism of the Prime Minister. The final budget for the project weighed in at close to £800 million, and further problems occurred when the government was tasked with selling the building after its closure, less than a year after opening. The building was eventually salvaged in 2005 when it was purchased under new plans to develop it into an entertainment hub. The result was the O₂ Arena, which opened in 2007.

The site of the former Pancras Gasworks is another example of a huge former industrial space that stood derelict for years, while it was decided what could be done to reuse it. Located directly behind Kings Cross and St Pancras railway stations, the land included several abandoned gasholder frames, train sheds, disused platforms and factories. It was a scene that contributed to the general opinion during the 1970s and '80s that Kings Cross was an unsavoury place, infamous for its problems with drugs and prostitution. It posed a significant problem for the London Borough of Camden, and the government as a whole, both of which were keen to attract investors that could boost the local economy and reverse decades of decline.

For a plot of land so closely related to London's railway network, it was fitting that what started the regeneration of the former gasworks and train yards was the arrival of a new railway line. Encouraged by government schemes aimed at attracting private companies to the site, it was decided in 2000 that St Pancras would become the new terminus station for High Speed 1 (HS1), the route of which includes the Eurostar service to France and Belgium. Construction of the new railway changed the face of the derelict land forever, and helped to attract other property developers. By 2006, plans had been approved for the entire area to be converted for new use as part of a project known as Kings Cross Central. Work commenced as soon as HS1 opened in 2007, and at time of writing several of the individual projects within the development scheme had already opened. It has brought life back to the area, and soon people will be living directly on top of the part of the site that was once home to the long lost gasworks. As mentioned in the Pancras Gasworks section earlier, the plan is for apartments to be built inside some of the original gasholder frames.

Over in west London, a similar tale to the one at Kings Cross had been unfolding in White City, on a site that was formerly both the Wood Lane Power Station and the Wood Lane Generating Station. Both plants had become derelict since closure, along with the disused Wood Lane London Underground station close by. It was an area of land that had become another victim of industrial decline in the city, and the local council struggled to find investors willing to redevelop the site.

An original proposal for a vast new shopping centre to be constructed fell through, because of financial issues suffered by several companies involved. Fortunately, it was at this point that major shopping centre company Westfield stepped in, and their multi-million pound mall opened here in 2008. It has been a phenomenal success, and, as explored earlier, the generating station building has been refurbished and worked into the new centre as a bus depot.

The Curse of Battersea Power Station

Although the former sites at Greenwich, White City and Kings Cross have finally managed to have their fortunes turned around, the same cannot be said about the iconic building that was once Battersea Power Station. It has been London's most infamous abandoned structure since its closure in the early 1980s, with numerous failed attempts at redevelopment that have left it standing as an impressive looking white elephant. What follows is a history of the most ambitious plans for Battersea that came and went without success.

The first major idea came a few years after closure, when it was announced that the power station was to become home to a new indoor theme park. Work began in 1987, followed by a high-profile press conference a year later, attended by Prime Minister Margaret Thatcher. It was during the early stages of construction that the roof was taken down so that the disused turbines and generators could be taken away.

The company responsible for the project ran into huge financial difficulties within two years of work starting. Costs had risen to a level almost £200 million over budget, and as a result the project was forced to shut down. By the early 1990s the theme park plan had been downgraded to a generic mixed-use project that was set to include offices and commercial floor space. This also hit problems, however, and the entire project had been abandoned by 1992.

The site was then purchased by a company based in Hong Kong. Their plan was to build a vast shopping centre, with restaurants, bars and a cinema. The

Decaying remains at Battersea Power Station.

once separate buildings (formerly power stations A and B) were even set to include different shops for different types of consumer. Spiralling costs also put an end to this project, once again leaving the power station in limbo.

Next to try and bring the building back to life was a property development company from Ireland. They purchased the site in 2006 and announced their intention to develop various different possible plans. The years that followed resulted in a mixed bag of ideas, none of which came to fruition. Some of the more conventional ideas were similar to those previously submitted by the company from Hong Kong, but with residential space also now added. The more offbeat proposals included turning the building into an eco-friendly village, with a section of the structure being reintroduced as a power station – albeit one that would be fuelled by biomass instead of the more heavily polluting coal or oil.

By 2011, none of the new proposals had begun to be built, and this company followed all those before it by abandoning their various Battersea Power Station projects once and for all. The primary limiting factor appeared once again to be a lack of budget, underlining just how much money it takes to redevelop a building on such an immense scale.

The fate of Battersea has provided the news media with years' worth of negative stories. When it was announced, therefore, in 2012 that the power station was up for sale again, it was viewed by many as being nothing more than the latest chapter in a long-running and ill-fated saga. The news was given some credence, however, when it transpired that one of the potential buyers happened to be Chelsea Football Club. In fact, the club had been involved with bids to purchase the site since 2008. Having outgrown their ground at Stamford Bridge since becoming one of the world's biggest teams, they planned to build a new stadium that incorporated elements of the famous building, including the iconic chimneys.

In the end Chelsea were unsuccessful in their bid, and instead the building was purchased by a property management company from Malaysia. At the time of writing, plans were being announced for work to begin as quickly as possible, with the aim being to turn the site into a mixture of entertainment, commercial and residential uses. The new project is also set to incorporate a much-rumoured extension to London Underground's Northern line, finally bringing the Tube network to an area of London that until now has always had to rely on crowded main-line train routes into central London.

It remains to be seen, of course, if the latest plan for Battersea Power Station will actually materialise, or if history is destined to repeat itself. Drawing similarities with former Conservative Prime Minister Margaret Thatcher and her support of the theme park idea, recent governments have been determined to see the old power station brought back into use. It is further proof of just how important the redevelopment of this famous building is, with several politicians and business leaders keen to show their full support.

Further down the river, Lots Road Power Station has shared a similar fate since it closed in 2002. A plan to convert the large building that remains for use as apartments was discussed in 2006, but caused controversy when it was announced that the blueprints for the project also included provision for two skyscrapers to be constructed next to the power station. The struggling economy appears to have halted all progress, however, and it is currently unclear who in fact owns Lots Road. For now it remains silent and abandoned, standing firm on some of the most desirable land in west London.

8

A DIRTY LEGACY FOR A CLEANER FUTURE

One of the reasons power stations and gasworks disappeared from the streets of London was the pollution they created. This was, however, more of a social debate during the 1950s and '60s, rather than one explicitly concerned with saving the planet. Living close to such sites and breathing in their fumes were conditions no longer tolerable for Londoners, and so the stations and works that survive today have been moved well away from the mass population.

The same demands for cleaner air quality and general well-being are more important than ever today, but Britain is also now a country more aware of the importance of protecting the environment. With a reduced carbon footprint at the forefront of many people's consciences, the government has been forced to set ambitious deadlines regarding improvements to the nation's environmental impact. In particular, it has committed to reducing carbon emissions.

As the largest and most important city in Britain, London has therefore had to lead by example, with many new projects and schemes being introduced to improve how we consume energy. The biggest challenge has been the one faced by industry, which has had to ditch older working methods of creating electricity and gas, and replace them with newer technologies that create less of a carbon footprint.

With population in London still on the rise, the challenge has been to maintain the high levels of power needed, without the negative effects on the environment. The outcome is a number of new developments across central and greater London, primarily concerned with sustainable electricity generation.

MODERN, GREENER POWER STATIONS

London's original power stations were built with little regard to the effect they would have on the environment. They were simply not the concerns of the day, and even if they were, the scientific research was not available to do much about it. There had been some basic concerns raised about the potentially damaging effects of industrial works as far back as the early nine-teenth century – decades before the first power stations arrived. Unlike today, however, where energy companies are more aware of the consequences of their actions, the thirst for huge profits up for grabs in supplying power to twentieth-century London proved to be a far more attractive prospect than any notion of saving the planet.

Every power station had its own set of chimneys, with some sites boasting several. They burned ton upon ton of fossil fuels every year, most fired by coal, with others later converted to oil. They created huge amounts of carbon dioxide emissions, polluting the air for thousands of people and contributing to the greenhouse effect in the process.

London no longer has any coal-fired power stations, and Greenwich is the only remaining example of an oil-fired plant. Instead, the remaining stations now tend to be operated by gas. Its lower carbon emission means it is a more environmentally friendly way of generating electricity, while still creating enough power to meet demand.

Most former power stations that were located in central London have disappeared without a trace, with redevelopment replacing them in many instances. But in Greater London, several plots of land selected on which to build original power stations are still suitable today, years after closure.

The lack of local residents and close proximity to water makes these former sites perennially desirable to power companies. It is for this reason that new gas-fired stations have been built close to the original plants at Barking and Brimsdown (the new site here is named Enfield Power Station). Taylors Lane in Willesden is also gas-fired, and Greenwich Power Station can be powered by gas if necessary (alongside oil, as discussed earlier).

In fact, Greenwich is a station with a long history of being able to move with the times. It started out as a coal-fired power station, but was then adapted to run its generators by oil. When Lots Road Power Station closed in 2002, it left Greenwich as sole back-up for the London Underground net-work. It warranted a series of updates, including conversion so that the boilers could be fired by either oil or gas.

In addition to carbon emissions, the other major issue highlighted today is the amount of energy that is often wasted during the electricity production process. Much energy is lost, for instance, when electricity is carried over long distances on its journey from power station to consumer, via miles of pylons, cables and substations. The further the current has to travel, the higher the wastage. It is therefore ironic that power stations were once moved further away from residential areas in order to improve people's quality of life, result-ing in the fact that electricity now has to travel much larger distances, creating lots of waste.

Taylors Lane Power Station.

Gas power stations may create fewer carbon emissions than coal or oil, but they do, however, still involve the release of toxins into the air. However, even the most dedicated environmental campaigner would agree that London needs electrical power on an enormous scale. The key therefore appears to be a balancing act between necessity and responsibility. If a power station cannot avoid a certain degree of carbon emission, then it should at least ensure that the emissions are reduced, and that there is as small an amount of energy wasted as possible.

It is a debate that has led to advancements in technology allowing for the construction of several 'combined cycle gas' power stations, of which Taylors Lane and Barking Reach are both examples. The principle is that wasted heat energy created during the process of generating electricity can be harnessed and used again as part of the actual process itself, usually as a way of fuelling the boilers. It is a cycle that greatly reduces the amount of wastage, which in turn also improves efficiency.

Gas and combined cycle gas power stations have therefore gone some way towards improving the image of the power industry. They still produce a substantial combined amount of carbon emissions, however, and that is why several new types of generating method have been explored in recent years. London leads the way in many of these new innovations, one of which is the concept of biomass power stations. It is a method that follows in the footsteps of the pioneering work of the Shoreditch Electric Light Station in east London, which used household waste to power its generators.

Waste matter, including household, industrial and even human waste, is today still used for power generation in many places around the world. One such operation can be found at Belvedere in Bexley, south-east London. It was a replacement for an earlier oil-fired power station on the same site. Most biomass power stations now tend to be operated by burning naturally occurring fuels, such as discarded plant life and wood.

A biomass power station was opened just outside Greater London in 2011, at Tilbury in Essex. There had been a power station at Tilbury since 1956, when a coal-fired plant was built here by the Central Electricity Board. It was decommissioned in 1981, having been replaced by a new coal-fired station in 1968, known as Tilbury 'B'. It was later converted to oil, followed by its conversion to biomass in 2011.

The new power station uses woodchip pellets imported from the USA and Canada, made out of trees from managed sustainable forests. Although it is only expected to operate for a short number of years, it is currently helping to supply electricity to well over a million homes in and around Greater London. The green credentials at the site at Tilbury were tarnished slightly in 2012, however, when a huge fire broke out that is said to have started in the bunkers used to store the woodchip pellets.

Research and feasibility studies are currently underway to select other sites in central and Greater London that would also be suitable for biomass power stations. Incredibly, if approved, some of the plans being discussed would see power generation return to several of London's abandoned power station and gasworks sites. The former Old Kent Road and Beckton gasworks have both been earmarked as potential choices. At Beckton, a new station would perhaps even see the coaling jetty brought back into action in order to handle delivery of biomass material.

Barking Reach Power Station, close to the original sites, has also been selected as a plant that could be converted in the same way as Tilbury, and Greenwich Power Station is also being considered for similar conversion. It would see Greenwich return to being fully functional (as opposed to simply being an emergency back-up for the London Underground). It would also allow for the large existing coal jetty to be reused.

SOLAR

Another obvious sustainable power source that London is using is the Sun. Solar panels have become a well-established method of generating electricity in homes, but the take-up on a larger scale has been slower than expected. In order for a building to be fully self-sufficient, enough roof top area space is needed to install the necessary panels. There is, of course, nowhere else in Britain with more large roof tops than London. A modest collection of high-profile companies have enhanced their buildings with solar power, including 8 Canada Square in the Docklands.

The use of solar panels has three major benefits. Firstly, they allow for the building's owner to cut overhead costs associated with electricity charges. Secondly, the company or individual gets to contribute towards a greener London. (Whether or not the reasons behind such a decision are motivated by a genuine sense of corporate responsibility, or simply for some positive PR is an issue sometimes brought into question.) Thirdly, any excess electricity provided by the solar panels can then be pumped back into the National Grid. The solar panels thus take some strain, however small, off the power stations, and therefore help contribute to lower emissions.

The largest solar power project in London can be found at Blackfriars station. One of the most historic Tube and main-line stations in the city, it has been rebuilt extensively in recent years. The platforms have been expanded so that they now run the length of a bridge across the Thames. It required construction of a new roof, on top of which over 4,000 solar panels have been installed. It is the largest solar-powered bridge in the world, and provides enough electricity to supply more than half of the station's power needs.

There are also private residential buildings contributing to greener London with their own solar panel installations. One such property is the 3 Acorns Retro Eco-House in Camberwell. It has been London's most environmentally friendly dwelling since 1997, with solar panels just one of many innovations that have helped the house to become carbon neutral.

WIND

Another natural resource used to generate electricity without harmful carbon emissions is wind. Land-based wind turbine farms require lots of open space, which makes finding suitable locations problematic in the urban sprawl of London. There are, however, a number of small farms around the city, including a site in Dagenham.

One of London's tallest buildings also generates its own power via the use of wind turbines. Known as Strata, it is located in Elephant & Castle, away from any other tall building. This makes it stand out for miles, helped by its three distinctive wind turbines at the top. It provides electricity for communal areas of the building, cutting energy bills for residents.

Similar to the prospects for London's future biomass power stations, a number of former electric power station and gasworks sites have also been put forward as potential large-scale wind farm locations. The Beckton and East Greenwich gasworks sites are both being considered, as is land close to the abandoned power stations at Barking, Brimsdown and Belvedere.

Space for wind turbines is far less restricted off-shore, however, and it is here, off the coast, where most of Britain's wind farms have been built so far. London's position on the banks of one of the most important rivers in the world makes it an ideal place to benefit from wind power. No turbines have been installed in the Thames as of yet, but a mega-sized wind farm was being constructed at the time of writing in the Thames Estuary, just off the coast of Kent. Known as London Array, it will include hundreds of turbines on completion. They will provide energy for a generating station in Kent, which can then be connected to the National Grid. With a greater output capacity than several of the lost power stations documented earlier, the farm has potential to supply hundreds of thousands of homes in and around Greater London.

In fact, although often controversial due to complaints over noise and the ruining of natural beauty spots, our island status means that wind power has the potential to supply much of Britain's electrical needs in generations to come.

NUCLEAR

The most controversial form of electricity generation today is nuclear power. The potential for catastrophic accidents means that all of Britain's nuclear power stations have been built well away from large population areas. This explains why London doesn't have one. The closest is located in Bradwell, Essex. It was one of seven nuclear power stations built by the Nuclear Power Group during the Central Electricity Generating Board era, and operated from 1962–2002. Although currently disused, there are plans being discussed to bring it back into operation in future years.

London did once have a connection to nuclear power, however. It was located not only in a highly populated area, but also one of its most historic and popular places. Known by the name JASON, it was a nuclear reactor used to supply power for the training staff who worked on nuclear submarine projects for the Navy. It was housed inside the Old Royal Naval College in Greenwich, and operated from 1962 until 1996.

REFERENCES & FURTHER READING

SELECTED BOOKS

Ackroyd, P., *London Under*. Vintage, 2012

Bayman, B., *Underground: Official Handbook*. Capital Transport, 2008

Czucha, F., *Old Sands End, Fulham*. Stenlake Publishing Limited, 2010

Emmerson, A., *The London Underground*. Shire Library, 2010

Francis, A., *Stepney Gasworks: The Archaeology and History of the Commercial Gas Light and Coke Company's works at Harford Street, London E1, 1837–1946*. Museum of London Archaeology, 2010

Halliday, S., *The Great Stink of London: Sir Joseph Bazalgette and the Cleansing of the Victorian Metropolis*. The History Press, 2009

Jones, R., *London's Transport: A Popular History*. Ian Allan, 2008

Kynaston, D., *Austerity Britain: 1945–1951*. Bloomsbury Publishing, 2007

Martin, A., *Underground Overground: A Passenger's History of the Tube*. Profile Books Ltd, 2012

Mathewson, A., Laval, D., Elton, J., Kentley, E. & Hulse, R., *The Brunels' Tunnel*. The Brunel Museum, 2006

Morrison, A. & Holder, G., *The History of Wood Lane, 1934–84*. BICC Research and Engineering, 1986

Sabbagh, K., *Power into Art: Creating the Tate Modern, Bankside*. Penguin Group, 2000

Stirling, E., *The History of The Gas Light & Coke Company 1812–1949*. A. & C. Black, 1992

Welbourn, N., *Lost Lines London*. Ian Allan, 2008

SELECTED PUBLICATIONS AND RESEARCH PAPERS

'A History of Haggerston Park'. Hackney City Farm, 2006

'Electricity Supply in the UK: A Chronology'. Electricity Council, 1987

'London Area Power Supply: A Survey of London's Electric Lighting and Power Stations', a self-published research article by Horne, M., 2012

'London Wind & Biomass Study Summary Report: Feasibility of the Potential for
    Stand Alone Wind and Biomass Plants in London'. London Energy Partnership,
    2006
'Powering Urban Transport' Information Sheet. London Transport Museum, 1994

SELECTED WEBSITES

Note: these websites were all available at the time of publication; they may since
    have lapsed.

http://carolineld.blogspot.co.uk/
http://greenwichindustrialhistory.blogspot.co.uk/
http://marysgasbook.blogspot.co.uk/
http://rbkclocalstudies.wordpress.com/
http://thevictorianist.blogspot.co.uk/
http://www.20thcenturylondon.org.uk/
http://www.aim25.ac.uk/http://www.biab.ac.uk/
http://www.british-history.ac.uk/
http://www.britishlistedbuildings.co.uk/
http://www.britishpathe.com/
http://www.crossness.org.uk/
http://www.engineering-timelines.com/
http://www.english-heritage.org.uk/
http://www.geograph.org.uk/
http://www.glias.org.uk/
http://www.hackneysociety.org/
http://www.ianvisits.co.uk/
http://www.londonrevolution.net/
http://www.ltmuseum.co.uk/
http://www.metadyne.co.uk/
http://www.museumoflondon.org.uk
http://www.nationalarchives.gov.uk/
http://www.newhamstory.com/
http://www.portcities.org.uk/
http://www.thegreenwichphantom.co.uk/
http://www.vauxhallcivicsociety.org.uk/

INDEX